HANDS-ON STANDARDS

*Photo-Illustrated Lessons for Teaching
with Math Manipulatives*

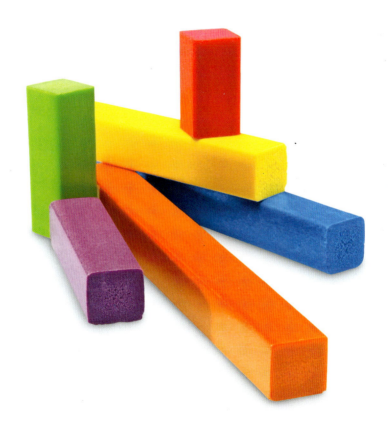

68 Hands-On Activities for
Grades 5–6

VERNON HILLS • KING'S LYNN

ISBN: 978-1-56911-273-1

LER 0853

380 N. Fairway Drive
Vernon Hills, IL 60061
(800) 222-3909

©Learning Resources, Inc., Vernon Hills, IL (U.S.A.)
Learning Resources Ltd., King's Lynn, Norfolk (U.K.)

All rights reserved. This book is copyrighted. No part of this book may be reproduced, stored in a retrieval system, or transmitted, in any form or by means electronic, mechanical, photocopying, recording, or otherwise, without written permission, except for the specific permission stated below.

Each blackline master or activity page is intended for reproduction in quantities sufficient for classroom use. Permission is granted to the purchaser to reproduce each blackline master or activity page in quantities suitable for noncommercial classroom use.

Printed in China.

Contents

Introduction		1
Research on the Benefits of Manipulatives		2
How to Use This Book		6
A Walk Through a Lesson		8
NCTM Correlation Chart		10

Number and Operations — 18

Lesson 1	Reason with Fractions	20
Lesson 2	Fractional Parts of a Collection	22
Lesson 3	Equivalent Fractions	24
Lesson 4	Decimals	26
Lesson 5	Equivalent Fractions and Decimals	28
Lesson 6	Compare and Order Fractions and Decimals	30
Lesson 7	Percents, Fractions, and Decimals	32
Lesson 8	Mixed Numbers, Decimals, and Percents Greater than 100%	34
Lesson 9	Factors, Primes, and Prime Factorization	36
Lesson 10	Squares and Square Roots	38
Lesson 11	Add Fractions with Unlike Denominators	40
Lesson 12	Subtract Fractions with Unlike Denominators	42
Lesson 13	Add and Subtract Decimals	44
Lesson 14	Multiply with Fractions	46
Lesson 15	Meaning of Division	48
Lesson 16	Divide with Fractions	50
Lesson 17	Multiply and Divide Decimals	52
Lesson 18	Ratios	54
Lesson 19	Proportions	56

Geometry — 58

Lesson 1	Measure and Classify Angles	60
Lesson 2	Identify and Classify Triangles	62
Lesson 3	Identify and Classify Quadrilaterals	64
Lesson 4	Regular Polygons	66
Lesson 5	Line Symmetry	68
Lesson 6	Parallel and Perpendicular Lines	70
Lesson 7	Shapes in the Coordinate Plane	72
Lesson 8	Slides and Flips	74
Lesson 9	Rotational Symmetry	76
Lesson 10	Multiple Transformations	78
Lesson 11	Tessellations	80
Lesson 12	Congruent Figures and Transformations	82
Lesson 13	Corresponding Parts of Congruent Figures	84
Lesson 14	Similar Triangles	86
Lesson 15	Nets	88
Lesson 16	Three-Dimensional Shapes	90

Algebra — 92

Lesson 1	Properties of Addition	94
Lesson 2	Properties of Multiplication	96
Lesson 3	Distributive Property	98
Lesson 4	Order of Operations	100
Lesson 5	Expressions with a Variable	102
Lesson 6	Equations with a Variable	104
Lesson 7	Addition and Subtraction Equations	106
Lesson 8	Multiplication and Division Equations	108

Lesson 9	Patterns and Function Tables	110
Lesson 10	Introduction to Integers	112
Lesson 11	Add Integers	114
Lesson 12	Subtract Integers	116
Lesson 13	Multiply Integers	118
Lesson 14	Divide Integers	120
Lesson 15	4-Quadrant Graphing	122
Lesson 16	Graphing Linear Equations	124

Measurement — 126

Lesson 1	Standard Units and Precision	128
Lesson 2	Perimeter and Area	130
Lesson 3	Area of a Parallelogram	132
Lesson 4	Area of a Triangle	134
Lesson 5	Surface Area of a Rectangular Solid	136
Lesson 6	Volume of a Rectangular Solid	138
Lesson 7	Volumes of Prisms and Pyramids	140
Lesson 8	Circumference of a Circle and π	142
Lesson 9	Area of a Circle	144

Data Analysis and Probability — 146

Lesson 1	Mean, Median, Mode, and Range	148
Lesson 2	Make a Conjecture Using a Scatterplot	150
Lesson 3	Line Graphs	152
Lesson 4	Circle Graphs	154
Lesson 5	Counting Principle	156
Lesson 6	Probability of an Event	158
Lesson 7	Complementary and Mutually Exclusive Events	160
Lesson 8	Probability of a Compound Event	162

Blackline Masters

BLM 1	Decimal Models	164
BLM 2	Fraction Circle in Hundredths	165
BLM 3	10 x 10 Grid	166
BLM 4	Tile Array	167
BLM 5	Hundred Chart	168
BLM 6	Inch Grid Paper	169
BLM 7	Eighths Fraction Squares	170
BLM 8	Centimeter Grid	171
BLM 9	Hundredths Grids	172
BLM 10	Function Table	173
BLM 11	1-Inch Number Lines	174
BLM 12	1-cm Number Lines	175
BLM 13	½-cm Number Lines	176
BLM 14	4-Quadrant Graph	177
BLM 15	Quadrilaterals Chart	178
BLM 16	Net Pattern	179
BLM 17	4-Column Recording Chart	180
BLM 18	6-Column Recording Chart	181
BLM 19	Line Measurement Worksheet	182

Glossary of Manipulatives 183

Index 186

Introduction

How can we make sure that students grasp the real meaning behind mathematical concepts instead of just memorizing numbers and repeating them back on tests? How can we help students develop an in-depth mathematical understanding?

Hands-On Standards: Photo-Illustrated Lessons for Teaching with Math Manipulatives (Grades 5–6) provides teachers with easy-to-access ways to help students "visualize mathematics." This 192-page manual delivers 69 age-appropriate lessons designed to engage teachers and students in meaningful, authentic learning. Each lesson defines the specific concepts and skills that students will be taught and includes step-by-step procedures for students to use in solving a problem that easily links math to their day-to-day lives. Full-color photographs highlight the steps used in hands-on learning. Because it is important to reinforce learning, each lesson provides additional ways to practice the concepts.

Hands-On Standards is divided into five sections—Number and Operations, Geometry, Algebra, Measurement, and Data Analysis and Probability. These sections are based on National Council of Teachers of Mathematics (NCTM) content strands. Each NCTM-focused lesson provides a structured framework for teachers to use manipulatives as tools to move students from the concrete to the abstract so that students can achieve understanding and succeed on standardized tests.

Each lesson in this book uses one of the following manipulatives:

- AngLegs™
- Base Ten Blocks
- Centimeter Cubes
- Color Tiles
- Cuisenaire® Rods
- Deluxe Rainbow Fraction® Circes
- Deluxe Rainbow Fraction® Squares
- Fraction Tower® Equivalency Cubes
- Geoboards
- GeoReflector™ Mirror
- Pattern Blocks
- Polyhedral Dice
- Rainbow Fraction® Rings
- Relational GeoSolids®
- Snap Cubes®
- Spinners
- Two-Color Counters

Research on the Benefits of Manipulatives

History of Manipulatives

Since ancient times, people of many different civilizations have used physical objects to help them solve everyday math problems. The ancient civilizations of Southwest Asia (the Middle East) used counting boards. These were wooden or clay trays covered with a thin layer of sand. The user would draw symbols in the sand to tally, for example, an account or take an inventory. The ancient Romans modified counting boards to create the world's first abacus. The Chinese abacus, which came into use centuries later, may have been an adaptation of the Roman abacus.

Similar devices were developed in the Americas. The Mayans and the Aztecs both had counting devices that featured corn kernels strung on string or wires stretched across a wooden frame. The Incas had their own unique counting tool—knotted strings called *quipu*.

The late 1800s saw the invention of the first true manipulatives—maneuverable objects that appeal to several different senses and are specifically designed for teaching mathematical concepts. Friedrich Froebel, a German educator who started the world's first kindergarten program in 1837, developed different types of objects to help his kindergarteners recognize patterns and appreciate geometric forms found in nature. In the early 1900s, Italian-born educator Maria Montessori further advanced the idea that manipulatives are important in education. She designed many materials to help preschool and elementary-school students discover and learn basic ideas in math and other subjects.

Since the early 1900s, manipulatives have come to be considered essential in teaching mathematics at the elementary-school level. In fact, for decades, the National Council of Teachers of Mathematics (NCTM) has recommended the use of manipulatives in teaching mathematical concepts at all grade levels (Hartshorn and Boren 1990).

Manipulatives and Curriculum Standards

The NCTM calls for manipulatives to be used in teaching a wide variety of topics in mathematics.

- sorting—a premathematical skill that aids in comprehension of patterns and functions
- ordering—a premathematical skill that enhances number sense and other math-related abilities
- distinguishing patterns—the foundation for making mathematical generalizations
- recognizing geometric shapes and understanding relationships among them
- making measurements, using both nonstandard and standard units with application to both two- and three-dimensional objects
- understanding the base-ten system of numbers
- comprehending mathematical operations—addition, subtraction, multiplication, division
- recognizing relationships among mathematical operations
- exploring and describing spatial relationships
- identifying and describing different types of symmetry
- developing and utilizing spatial memory
- learning about and experimenting with transformations
- engaging in problem solving
- representing mathematical ideas in a variety of ways
- connecting different concepts in mathematics
- communicating mathematical ideas effectively

Different states across the nation have also mandated the use of manipulatives for teaching math. These include California, North Carolina, Texas, and Tennessee, among others. In addition, many local school districts mandate or strongly suggest that manipulatives be used in teaching math, especially at the elementary level.

Manipulative use is recommended because it is supported by both learning theory and educational research in the classroom.

Concrete stage	Representational stage	Abstract stage
A mathematical concept is introduced with manipulatives; students explore the concept using the manipulatives in purposeful activity.	A mathematical concept is represented using pictures of some sort to stand for the concrete objects (the manipulatives) of the previous stage; students demonstrate how they can both visualize and communicate the concept at a pictorial level.	Mathematical symbols (numerals, operation signs, etc.) are used to express the concept in symbolic language; students demonstrate their understanding of the mathematical concept using the language of mathematics.

How Learning Theory Supports the Use of Manipulatives

The theory of experiential education revolves around the idea that learning is enhanced when students acquire knowledge through active processes that engage them (Hartshorn and Boren 1990). Manipulatives can be key in providing effective, active, engaging lessons in the teaching of mathematics.

Manipulatives help students learn by allowing them to move from concrete experiences to abstract reasoning (Heddens 1986; Reisman 1982; Ross and Kurtz 1993). This strategy is exemplified by the Concrete-Representational-Abstract (CRA) instructional approach. The three-stage CRA process is summarized above (The Access Center, October 1, 2004).

The use of manipulatives helps students hone their mathematical thinking skills. According to Stein and Bovalino (2001), "Manipulatives can be important tools in helping students to think and reason in more meaningful ways. By giving students concrete ways to compare and operate on quantities, such manipulatives as pattern blocks, tiles, and cubes can contribute to the development of well-grounded, interconnected understandings of mathematical ideas."

To gain a deep understanding of mathematical ideas, students need to be able to integrate and connect a variety of concepts in many different ways. Clements (1999) calls this type of deep understanding "Integrated-Concrete" knowledge. The effective use of manipulatives can help students connect ideas and integrate their knowledge so that they gain a deep understanding of mathematical concepts.

Teachers play a crucial role in helping students use manipulatives successfully, so that they move through the three stages of learning and arrive at a deep understanding of mathematical concepts.

How Research from the Classroom Supports the Use of Manipulatives

Over the past four decades, studies done at all different grade levels and in several different countries indicate that mathematical achievement increases when manipulatives are put to good use (Canny 1984; Clements 1999; Clements and Battista 1990; Dienes 1960; Driscoll 1981; Fennema 1972, 1973; Skemp 1987; Sugiyama 1987; Suydam 1984). Additional research shows that use of manipulatives over the long-term provides more benefits than short-term use does (Sowell 1989).

With long-term use of manipulatives in mathematics, educators have found that students make gains in the following general areas (Heddens 1986; Picciotto 1998; Sebesta and Martin 2004):

- verbalizing mathematical thinking
- discussing mathematical ideas and concepts
- relating real-world situations to mathematical symbolism
- working collaboratively
- thinking divergently to find a variety of ways to solve problems
- expressing problems and solutions using a variety of mathematical symbols
- making presentations
- taking ownership of their learning experiences
- gaining confidence in their abilities to find solutions to mathematical problems

Studies have shown that students using manipulatives in specific mathematical subjects are more likely to achieve success than students who don't have the opportunity to work with manipulatives. The following are some specific areas in which research shows manipulatives are especially helpful:

Counting Some students need to use manipulatives to learn to count (Clements 1999).

Place Value Using manipulatives increases students' understanding of place value (Phillips 1989).

Computation Students learning computational skills tend to master and retain these skills more fully when manipulatives are used as part of their instruction (Carroll and Porter 1997).

Problem Solving Using manipulatives has been shown to help students reduce errors and increase their scores on tests that require them to solve problems (Carroll and Porter 1997; Clements 1999; Krach 1998).

Fractions Students who use appropriate manipulatives to help learn fractions outperform students who rely only on textbooks when tested on these concepts (Jordan, Miller, and Mercer 1998; Sebesta and Martin 2004).

Ratios Students who use appropriate manipulatives to help them learn fractions also have significantly improved achievement when tested on ratios when compared to students who do not have exposure to these manipulatives (Jordan, Miller, and Mercer 1998).

Algebraic Abilities Algebraic abilities include the ability to represent algebraic expressions, to interpret such expressions, to make connections between concepts when solving linear equations, and to communicate algebraic concepts. Research indicates that students who used manipulatives in their mathematics classes have higher algebraic abilities than those who did not use manipulatives (Chappell and Strutchens 2001).

Manipulatives have also been shown to provide a strong foundation for students mastering the following mathematical concepts (The Access Center, October 1, 2004):

- number relations
- measurement
- decimals
- number bases
- percentages
- probability
- statistics

Well-known math educator Marilyn Burns (Burns 2005) considers manipulatives essential for teaching math to students of all levels. She finds that manipulatives help make math concepts accessible to almost all learners, while at the same time offering ample opportunities to challenge students who catch on quickly to the concepts being taught. Research indicates that using manipulatives is especially useful for teaching low achievers, students with learning disabilities, and English language learners (Marsh and Cooke 1996; Ruzic and O'Connell 2001).

Research also indicates that using manipulatives helps improve the environment in math classrooms. When students work with manipulatives and are then given a chance to reflect on their experiences, not only is mathematical learning enhanced, math anxiety is greatly reduced (Cain-Caston 1996; Heuser 2000). Exploring manipulatives, especially in a self-directed manner, provides an exciting classroom environment and promotes in students a positive attitude toward learning (Heuser 1999; Moch 2001). Among the benefits several researchers found for using manipulatives was that they helped make learning fun (Moch 2001; Smith et al. 1999).

Summary

Research from both learning theory and classroom studies shows that using manipulatives to help teach math can positively affect learning. This is true for students at all levels and of all abilities. It is also true for almost every topic covered in elementary-school mathematics curricula. Papert (1980) calls manipulatives "objects to think with." Incorporating manipulatives into mathematics lessons in meaningful ways helps students grasp concepts with greater ease, making teaching most effective.

Reference Citations

The Access Center, http://coe.jmu.edu/mathvidsr/disabilities.htm (October 1, 2004)

Burns, M. (1996). How to make the most of math manipulatives. *Instructor,* accessed at http://teacher.scholastic.com/lessonrepro/lessonplans/instructor/burns.htm.

Cain-Caston, M. (1996). Manipulative queen. *Journal of Instructional Psychology,* 23(4): 270–274.

Canny, M. E. (1984). The relationship of manipulative materials to achievement in three areas of fourth-grade mathematics: Computation, concept development, and problem solving. *Dissertation Abstracts International,* 45A: 775–776.

Carroll, W. M. & Porter, D. (1997). Invented strategies can develop meaningful mathematical procedures. *Teaching Children Mathematics,* 3(7): 370–374.

Chappell, M. F. & Strutchens, M. E. (2001). Creating connections: Promoting algebraic thinking with concrete models. *Mathematics Teaching in the Middle School.* Reston, VA: National Council of Teachers of Mathematics.

Clements, D. H. (1999). "Concrete" manipulatives, concrete ideas. *Contemporary Issues in Early Childhood,* 1(1): 45–60.

Clements, D. H. & Battista, M. T. (1990). Constructive learning and teaching. *The Arithmetic Teacher,* 38: 34–35.

Dienes, Z. P. (1960). *Building up mathematics.* London: Hutchinson Educational.

Driscoll, M. J. (1984). What research says. *The Arithmetic Teacher,* 31: 34–35.

Fennema, E. H. (1972). Models and mathematics. *The Arithmetic Teacher,* 19: 635–640.

———. (1973). Manipulatives in the classroom. *The Arithmetic Teacher,* 20: 350–352.

Hartshorn, R. & Boren, S. (1990). Experiential learning of mathematics: Using manipulatives. *ERIC Clearinghouse on Rural Education and Small Schools.*

Heddens, J. W. (1986). Bridging the gap between the concrete and the abstract. *The Arithmetic Teacher,* 33: 14–17.

———. Improving mathematics teaching by using manipulatives. Kent State University, accessed at www.fed.cubk.edu.hk.

Heuser, D. (1999). Reflections on teacher philosophies and teaching strategies upon children's cognitive structure development—reflection II; Pennsylvania State University, accessed at http://www.ed.psu.edu/CI/Journals/1999AETS/Heuser.rtf

———. (2000). Mathematics class becomes learner centered. *Teaching Children Mathematics,* 6(5): 288–295.

Jordan, L., Miller, M., & Mercer, C. D. (1998). The effects of concrete to semi-concrete to abstract instruction in the acquisition and retention of fraction concepts and skills. *Learning Disabilities: A Multidisciplinary Journal,* 9: 115–122.

Krach, M. (1998). Teaching fractions using manipulatives. *Ohio Council of Teachers of Mathematics,* 37: 16–23.

Maccini, P. & Gagnon, J. A. (2000, January). Best practices for teaching mathematics to secondary students with special needs. *Focus on Exceptional Children,* 32 (5): 11.

Marsh, L. G. & Cooke, N. L. (1996). The effects of using manipulatives in teaching math problem solving to students with learning disabilities. *Learning Disabilities Research & Practice,* 11(1): 58–65.

Martino, A. M. & Maher, C. A. (1999). Teacher questioning to promote justification and generalization in mathematics: What research practice has taught us. *Journal of Mathematical Behavior,* 18(1): 53–78.

Moch, P. L. (Fall 2001). Manipulatives work! *The Educational Forum.*

Nunley, K. F. (1999). *Why hands-on tasks are good.* Salt Lake City, UT: Layered Curriculum.

Papert, S. (1980). *Mindstorms.* Scranton, PA: Basic Books.

Phillips, D. G. (1989) The development of logical thinking: A three-year longitudinal study. Paper presented to the National Council of Teachers of Mathematics, Orlando, FL.

Picciotto, H. (1998). Operation sense, tool-based pedagogy, curricular breadth: a proposal, accessed at http://www.picciotto.org.

Pugalee, D. K. (1999). Constructing a model of mathematical literacy. *The Clearing House* 73(1): 19–22.

Reisman, F. K. (1982). *A guide to the diagnostic teaching of arithmetic* (3rd ed.). Columbus, OH: Merrill.

Ross, R. & Kurtz, R. (1993). Making manipulatives work: A strategy for success. *The Arithmetic Teacher* (January 1993). 40: 254–258.

Ruzic, R. & O'Connell, K. (2001). Manipulatives. *Enhancement Literature Review,* accessed at http://www.cast.org/ncac/Manipulatives1666.cfm.

Sebesta, L. M. & Martin, S. R. M. (2004). Fractions: building a foundation with concrete manipulatives. *Illinois Schools Journal,* 83(2): 3–23.

Skemp, R. R. (1987). *The psychology of teaching mathematics* (revised American edition). Hillsdale, NJ: Erlbaum.

Smith, N. L., Babione, C., & Vick, B. J. (1999). Dumpling soup: Exploring kitchens, cultures, and mathematics. *Teaching Children Mathematics,* 6: 148–152.

Sowell, E. (1989). Effects of manipulative materials in mathematics instruction. *Journal for Research in Mathematics Education,* 20: 498–505.

Stein, M. K. & Bovalino, J. W. (2001). Manipulatives: One piece of the puzzle. *Mathematics Teaching in Middle School,* 6(6): 356–360.

Sugiyama, Y. (1987). Comparison of word problems in textbooks between Japan and the U.S. in J. P. Becker & T. Miwa (eds.), *Proceedings of U.S.–Japan Seminar on Problem Solving.* Carbondale, IL: Board of Trustees, Southern Illinois University.

Suydam, M. (1984). Research report: manipulative materials. *The Arithmetic Teacher,* 31: 27.

How to Use This Book

The goal of *Hands-On Standards* is to transition students from informal, concrete strategies to more formal, abstract ones. This book is based on the use of manipulatives, which are perfect tools for teaching and reinforcing learning. Manipulatives

- are meaningful to students;
- provide students with control and flexibility;
- mirror cognitive and mathematical structures in meaningful ways; and
- help students in connecting different types of knowledge.

Built around the manipulatives are activities that engage memory so that students more readily retain the mathematical concepts they learn.

The First Step

Before even opening the book, create a learning environment in which your students are excited about embarking on their mathematics adventure.

- First, take time to become familiar with the manipulatives and the lessons in which they are used.
- The lessons have been written to be used with common manipulatives. However, depending on the resources available, you may need to substitute one manipulative with another.
- Introduce your students to the manipulatives they will encounter. Have them investigate the manipulative kits in an unstructured way so that they will become familiar with all of the manipulative shapes and textures.
- Allow your students the time to discover the relationships between sizes and shapes. Let them have fun.
- Keep the manipulative kits in a special place. Make sure students know where the manipulatives are stored so that they can easily access them during math time and classroom free time.

Getting Ready

Once students have had time to use the manipulatives, walk them through a sample lesson.

- You might want to model an activity so that students can see how you use the manipulatives.
- Make sure students know that there is a trial-and-error process that they must go through so that they aren't self-conscious if they make errors.
- Tell students that you will all talk about the activities afterward and that they will be able to write about the activities as well.

Using the Manual

The lessons in *Hands-On Standards* have been organized so that you can make an easy progression through the book. However, feel free to teach the lessons in any order to maximize students' learning. Following is a suggested plan for teaching each lesson:

- Read the story problem to students. Ask them if they have ever had a similar problem. Let them tell you their experiences.

- Define any necessary vocabulary. Give students ways to use the words so they become familiar with the concepts.

- Divide the class into groups or pairs, depending on the directions. Show students the manipulatives they will be using. Give them a few minutes to get their supplies ready.

- While the lessons have been designed for use with individual students, pairs, or small groups, they can easily be adapted to meet your own classroom organization or teaching preference. Lessons can be used in centers or in a more traditional classroom setting.

- Lessons have been written for a fairly broad age range, for example, grades 3–4. While the lessons serve as a guide, you should feel free to adapt data, vocabulary, and complexity to what you consider developmentally appropriate for your students.

- Help students perform the **Try It!** activity. Make sure they are having success as they work to understand the concept and develop an answer to the problem.

- Discuss the activity with students when they are finished. For suggestions, see the **Talk About It** section.

- Ask students to follow up the discussion by using the prompt in the **Solve It** section.

- Finally, have students work the problem in the **Standardized Practice** section without using manipulatives.

However you decide to use these lessons, make this manual your own. Use the ideas as jumping off points to enhance your teaching style and your existing math curriculum.

A Walk Through a Lesson

Each lesson in *Hands-On Standards* includes many features, including background information, objectives, pacing and grouping suggestions, discussion questions, and ideas for further activities, all in addition to the step-by-step, hands-on activity instruction. Take a walk through a lesson to see an explanation of each feature.

Lesson Introduction
A brief introduction explores the background of the concepts and skills covered in each lesson. It shows how they fit into the larger context of students' mathematical development.

Try It! Arrow
In order to provide a transition from the introduction to the activity, an arrow draws attention to the Try It! activity on the next page. When the activity has been completed, return to the first page to complete the lesson.

Objective
The **Objective** summarizes the skill or concept students will learn through the hands-on lesson.

Skills
The **Skills** box lists the top three mathematical skills that students will use in each lesson.

NCTM Expectations
Each lesson has been created to align with one or more of the grade-level expectations set by the National Council of Teachers of Mathematics (NCTM) in their *Principles and Standards for School Mathematics* (2000).

Talk About It
The **Talk About It** section provides post-activity discussion topics and questions. Discussion reinforces activity concepts and provides the opportunity to make sure students have learned and understood the concepts and skills.

Solve It
Solve It gives students a chance to show what they've learned. Students are asked to return to and solve the original word problem. They might summarize the lesson concept through drawing or writing, or extend the skill through a new variation on the problem.

More Ideas
More Ideas provides additional activities and suggestions for teaching about the lesson concept using a variety of manipulatives. These ideas might be suggestions for additional practice with the skill or an extension of the lesson.

Standardized Practice
Standardized Practice allows students to confront the lesson concept as they might encounter it on a standardized test.

LESSON 1

Number and Operations

Reason with Fractions

The understanding of a fraction as part of a whole is essential to all other work with fractions. By using various area models for parts of a whole, students see how fractions relate to the whole and to each other. Here, students are encouraged to think flexibly to compare fractions. They apply their understanding of fractions to reason about the amount that a fraction represents.

Try It! Perform the Try It! activity on the next page.

Objective
Reason with fractions by comparing parts of a whole.

Skills
- Representing rational numbers
- Comparing fractions
- Reasoning

NCTM Expectations
Grades 3–5
Number and Operations
- Develop understanding of fractions as parts of unit wholes, as parts of a collection, as locations on number lines, and as divisions of whole numbers.
- Use models, benchmarks, and equivalent forms to judge the size of fractions.

Talk About It
Discuss the Try It! activity.
- **Ask:** Which is smaller, $\frac{1}{6}$ of a pie or $\frac{1}{8}$ of a pie?
- **Ask:** Which is greater, $\frac{5}{6}$ of a pie or $\frac{7}{8}$ of a pie?
- **Ask:** How are your answers to these two questions related?

Solve It
Reread the problem with students. Students should reason that since the slice taken from the blueberry pie is smaller, more blueberry pie remains. Each pie has one fewer slice, but the removed slices were not the same size. Have students write the inequalities used to solve this problem: $\frac{1}{8} < \frac{1}{6}$ and $\frac{7}{8} > \frac{5}{6}$.

More Ideas
For other ways to teach about reasoning with fractions—
- Use fraction squares instead of fraction circles to solve this problem. Compare the rectangles that represent $\frac{1}{6}$ and $\frac{1}{8}$.
- Using Cuisenaire® Rods, tell students that each of the orange rods represents one whole. Ask students which rods represent $\frac{1}{2}$. Have students demonstrate their answers by laying two of these rods beneath an orange rod. Repeat this approach for $\frac{1}{5}$ and $\frac{1}{10}$. Ask students to compare the rods that represent $\frac{1}{2}$, $\frac{1}{5}$, and $\frac{1}{10}$.

Standardized Practice
Have students try the following problem.

Which shows the fractions listed in order from least to greatest?

A. $\frac{1}{5}, \frac{1}{6}, \frac{4}{5}, \frac{5}{6}$

B. $\frac{1}{6}, \frac{1}{5}, \frac{4}{5}, \frac{5}{6}$

C. $\frac{1}{6}, \frac{1}{5}, \frac{5}{6}, \frac{4}{5}$

D. $\frac{1}{6}, \frac{5}{6}, \frac{1}{5}, \frac{4}{5}$

20

Try It!

The **Try It!** activity opens with **Pacing** and **Grouping** guides. The **Pacing** guide indicates about how much time it will take for students to complete the activity, including the post-activity discussion. The **Grouping** guide recommends whether students should work independently, in pairs, or in small groups.

Next, the **Try It!** activity is introduced with a real-world story problem. Students will "solve" the problem by performing the hands-on activity. The word problem provides a context for the hands-on work and the lesson skill.

The **Materials** box lists the type and quantity of materials that students will use to complete the activity, including manipulatives such as Color Tiles and Pattern Blocks.

This section of the page also includes any instruction that students may benefit from before starting the activity, such as a review of foundational mathematical concepts or an introduction to new ones.

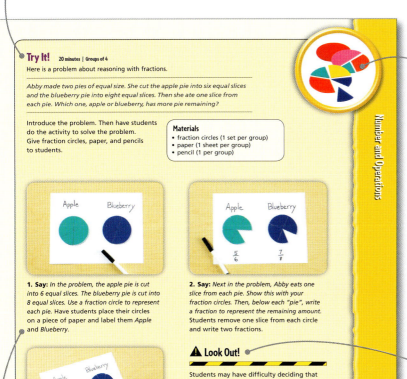

A thumbnail image quickly identifies the main manipulative used in the Try It! activity.

Look Out!
Look Out! describes common errors or misconceptions likely to be exhibited by students at this age dealing with each skill or concept and offers troubleshooting suggestions.

Step-by-Step Activity Procedure
The hands-on activity itself is the core of each lesson. It is presented in three—or sometimes four—steps, each of which includes instruction in how students should use manipulatives and other materials to address the introductory word problem and master the lesson's skill or concept. An accompanying photograph illustrates each step.

NCTM Correlation Chart

*NCTM Standards/Grades 3-5 Expectations	Lessons
Number and Operations	
Understand numbers, ways of representing numbers, relationships among numbers, and number systems.	
Understand the place-value structure of the base-ten number system and be able to represent and compare whole numbers and decimals.	NUM-4
Recognize equivalent representations for the same number and generate them by decomposing and composing numbers.	NUM-3
Develop understanding of fractions as parts of unit wholes, as parts of a collection, as locations on number lines, and as divisions of whole numbers.	NUM-1; NUM-2; NUM-5
Use models, benchmarks, and equivalent forms to judge the size of fractions.	NUM-1; NUM-3; NUM-5
Recognize and generate equivalent forms of commonly used fractions, decimals, and percents.	NUM-5; NUM-7; NUM-11; NUM-12
Explore numbers less than 0 by extending the number line and through familiar applications.	
Describe classes of numbers according to characteristics such as the nature of their factors.	
Understand meanings of operations and how they relate to one another.	
Understand various meanings of multiplication and division.	NUM-15
Understand the effects of multiplying and dividing whole numbers.	NUM-15
Identify and use relationships between operations, such as division as the inverse of multiplication, to solve problems.	NUM-15; ALG-7; ALG-8
Understand and use properties of operations, such as the distributivity of multiplication over addition.	ALG-4
Compute fluently and make reasonable estimates.	
Develop fluency with basic number combinations for multiplication and division and use these combinations to mentally compute related problems, such as 30 × 50.	
Develop fluency in adding, subtracting, multiplying, and dividing whole numbers.	NUM-15; ALG-6; ALG-7; ALG-8
Develop and use strategies to estimate the results of whole-number computations and to judge the reasonableness of such results.	
Develop and use strategies to estimate computations involving fractions and decimals in situations relevant to students' experience.	
Use visual models, benchmarks, and equivalent forms to add and subtract commonly used fractions and decimals.	NUM-11; NUM-12

*Reprinted with permission from *Principles and Standards for School Mathematics,* copyright 2000 by the National Council of Teachers of Mathematics. All rights reserved. Standards are listed with the permission of the National Council of Teachers of Mathematics (NCTM). NCTM does not endorse the content or validity of these alignments.

Number and Operations Continued

Select appropriate methods and tools for computing with whole numbers from among mental computation, estimation, calculators, and paper and pencil according to the context and nature of the computation and use the selected method or tools.	

*NCTM Standards/Grades 3-5 Expectations	Lessons
Geometry	
Analyze characteristics and properties of two- and three-dimensional geometric shapes and develop mathematical arguments about geometric relationships.	
Identify, compare, and analyze attributes of two- and three-dimensional shapes and develop vocabulary to describe the attributes.	GEO-1; GEO-2; GEO-3; GEO-12; GEO-15;
Classify two- and three-dimensional shapes according to their properties and develop definitions of classes and shapes such as triangles and pyramids.	GEO-2; GEO-3
Investigate, describe, and reason about the results of subdividing, combining, and transforming shapes.	GEO-11; GEO-15
Explore congruence and similarity.	GEO-12; GEO-14
Make and test conjectures about geometric properties and relationships and develop logical arguments to justify conclusions.	GEO-11
Specify locations and describe spatial relationships using coordinate geometry and other representational systems.	
Describe location and movement using common language and geometric vocabulary.	
Make and use coordinate systems to specify locations and to describe paths.	ALG-15
Find the distance between points along horizontal and vertical lines of a coordinate system.	
Apply transformations and use symmetry to analyze mathematical situations.	
Predict and describe the results of sliding, flipping, and turning two-dimensional shapes.	
Describe a motion or a series of motions that will show that two shapes are congruent.	GEO-12
Identify and describe line and rotational symmetry in two- and three-dimensional shapes and designs.	GEO-5
Use visualization, spatial reasoning, and geometric modeling to solve problems.	
Build and draw geometric objects.	GEO-1; GEO-2; GEO-3; GEO-5; GEO-14; GEO-15
Create and describe mental images of objects, patterns, and paths.	GEO-5

*Reprinted with permission from *Principles and Standards for School Mathematics*, copyright 2000 by the National Council of Teachers of Mathematics. All rights reserved. Standards are listed with the permission of the National Council of Teachers of Mathematics (NCTM). NCTM does not endorse the content or validity of these alignments.

Geometry Continued

Identify and build a three-dimensional object from two-dimensional representations of that object.	GEO-15
Identify and build a two-dimensional representation of a three-dimensional object.	
Use geometric models to solve problems in other areas of mathematics, such as number and measurement.	
Recognize geometric ideas and relationships and apply them to other disciplines and to problems that arise in the classroom or in everyday life.	GEO-1; GEO-5; GEO-12; GEO-14

*NCTM Standards/Grades 3–5 Expectations	Lessons
Algebra	
Understand patterns, relations, and functions.	
Describe, extend, and make generalizations about geometric and numeric patterns.	ALG-9
Represent and analyze patterns and functions, using words, tables, and graphs.	ALG-9
Represent and analyze mathematical situations and structures using algebraic symbols.	
Identify such properties as commutativity, associativity, and distributivity and use them to compute with whole numbers.	ALG-1; ALG-2; ALG-3
Represent the idea of a variable as an unknown quantity using a letter or a symbol.	ALG-6
Express mathematical relationships using equations.	ALG-1; ALG-2; ALG-3; ALG-6; ALG-7; ALG-8
Use mathematical models to represent and understand quantitative relationships.	
Model problem situations with objects and use representations such as graphs, tables, and equations to draw conclusions.	ALG-1; ALG-2; ALG-3; ALG-4; ALG-6; ALG-7; ALG-8; ALG-9
Analyze change in various contexts.	
Investigate how a change in one variable relates to a change in a second variable.	
Identify and describe situations with constant or varying rates of change and compare them.	

*NCTM Standards/Grades 3–5 Expectations	Lessons
Measurement	
Understand measurable attributes of objects and the units, systems, and processes of measurement.	
Understand such attributes as length, area, weight, volume, and size of angle and select the appropriate type of unit for measuring each attribute.	MEA-2; MEA-4
Understand the need for measuring with standard units and become familiar with standard units in the customary and metric systems.	MEA-1
Carry out simple unit conversions, such as from centimeters to meters, within a system of measurement.	

*Reprinted with permission from *Principles and Standards for School Mathematics*, copyright 2000 by the National Council of Teachers of Mathematics. All rights reserved. Standards are listed with the permission of the National Council of Teachers of Mathematics (NCTM). NCTM does not endorse the content or validity of these alignments.

Measurement Continued

Understand that measurements are approximations and understand how differences in units affect precision.	MEA-1
Explore what happens to measurements of a two-dimensional shape such as its perimeter and area when the shape is changed in some way.	MEA-2
Apply appropriate techniques, tools, and formulas to determine measurements.	
Develop strategies for estimating the perimeters, areas, and volumes of irregular shapes.	
Select and apply appropriate standard units and tools to measure length, area, volume, weight, time, temperature, and the size of angles.	MEA-1
Select and use benchmarks to estimate measurements.	
Develop, understand, and use formulas to find the area of rectangles and related triangles and parallelograms.	MEA-2; MEA-3; MEA-4
Develop strategies to determine the surface areas and volumes of rectangular solids.	

*NCTM Standards/Grades 3–5 Expectations	Lessons
Data Analysis and Probability	
Formulate questions that can be addressed with data and collect, organize, and display relevant data to answer them.	
Design investigations to address a question and consider how data-collection methods affect the nature of the data set.	
Collect data using observations, surveys, and experiments.	
Represent data using tables and graphs such as line plots, bar graphs, and line graphs.	DAT-3
Recognize the differences in representing categorical and numerical data.	
Select and use appropriate statistical methods to analyze data.	
Describe the shape and important features of a set of data and compare related data sets, with an emphasis on how the data are distributed.	
Use measures of center, focusing on the median, and understand what each does and does not indicate about the data set.	
Compare different representations of the same data and evaluate how well each representation shows important aspects of the data.	
Develop and evaluate inferences and predictions that are based on data.	
Propose and justify conclusions and predictions that are based on data and design studies to further investigate the conclusions or predictions.	DAT-3
Understand and apply basic concepts of probability.	
Describe events as likely or unlikely and discuss the degree of likelihood using such words as *certain*, *equally likely*, and *impossible*.	
Predict the probability of outcomes of simple experiments and test the predictions.	
Understand that the measure of the likelihood of an event can be represented by a number from 0 to 1.	

*Reprinted with permission from *Principles and Standards for School Mathematics*, copyright 2000 by the National Council of Teachers of Mathematics. All rights reserved. Standards are listed with the permission of the National Council of Teachers of Mathematics (NCTM). NCTM does not endorse the content or validity of these alignments.

*NCTM Standards/Grades 6-8 Expectations	Lessons
Number and Operations	
Understand numbers, ways of representing numbers, relationships among numbers, and number systems.	
Work flexibly with fractions, decimals, and percents to solve problems.	NUM-6; NUM-8; NUM-14; NUM-16; NUM-17
Compare and order fractions, decimals, and percents efficiently and find their approximate locations on a number line.	NUM-6; NUM-8
Develop meaning for percents greater than 100 and less than 1.	NUM-8
Understand and use ratios and proportions to represent quantitative relationships.	NUM-18; NUM-19
Develop an understanding of large numbers and recognize and appropriately use exponential, scientific, and calculator notation.	
Use factors, multiples, prime factorization, and relatively prime numbers to solve problems.	NUM-9
Develop meaning for integers and represent and compare quantities with them.	ALG-10
Understand meanings of operations and how they relate to one another.	
Understand the meaning and effects of arithmetic operations with fractions, decimals, and integers.	NUM-13; NUM-14; NUM-16; NUM-17; ALG-11; ALG-12; ALG-13; ALG-14
Use the associative and commutative properties of addition and multiplication and the distributive property of multiplication over addition to simplify computations with integers, fractions, and decimals.	
Understand and use the inverse relationships of addition and subtraction, multiplication and division, and squaring and finding square roots to simplify computations and solve problems.	NUM-10; NUM-13; ALG-12; ALG-14
Compute fluently and make reasonable estimates.	
Select appropriate methods and tools for computing with fractions and decimals from among mental computation, estimation, calculators, and paper and pencil, depending on the situation, and apply the selected methods.	
Develop and analyze algorithms for computing with fractions, decimals, and integers and develop fluency in their use.	ALG-11; ALG-12; ALG-13; ALG-14
Develop and use strategies to estimate the results of rational-number computations and judge the reasonableness of the results.	
Develop, analyze, and explain methods for solving problems involving proportions, such as scaling and finding equivalent ratios.	NUM-19

*Reprinted with permission from *Principles and Standards for School Mathematics*, copyright 2000 by the National Council of Teachers of Mathematics. All rights reserved. Standards are listed with the permission of the National Council of Teachers of Mathematics (NCTM). NCTM does not endorse the content or validity of these alignments.

*NCTM Standards/Grades 6-8 Expectations	Lessons
Geometry	
Analyze characteristics and properties of two- and three-dimensional geometric shapes and develop mathematical arguments about geometric relationships.	
Precisely describe, classify, and understand relationships among types of two- and three-dimensional objects using their defining properties.	GEO-4; GEO-13; GEO-16
Understand relationships among the angles, side lengths, perimeters, areas, and volumes of similar objects.	
Create and critique inductive and deductive arguments concerning geometric ideas and relationships, such as congruence, similarity, and the Pythagorean relationship.	GEO-13
Specify locations and describe spatial relationships using coordinate geometry and other representational systems.	
Use coordinate geometry to represent and examine the properties of geometric shapes.	GEO-6; GEO-7; GEO-8
Use coordinate geometry to examine special geometric shapes, such as regular polygons or those with pairs of parallel or perpendicular sides.	GEO-6; GEO-7
Apply transformations and use symmetry to analyze mathematical situations.	
Describe sizes, positions, and orientations of shapes under informal transformations such as flips, turns, slides, and scaling.	GEO-8; GEO-9; GEO-10; GEO-13
Examine the congruence, similarity, and line or rotational symmetry of objects using transformations.	GEO-9
Use visualization, spatial reasoning, and geometric modeling to solve problems.	
Draw geometric objects with specified properties, such as side lengths or angle measures	GEO-4; GEO-7
Use two-dimensional representations of three-dimensional objects to visualize and solve problems such as those involving surface area and volume.	
Use visual tools such as networks to represent and solve problems.	
Use geometric models to represent and explain numerical and algebraic relationships.	
Recognize and apply geometric ideas and relationships in areas outside the mathematics classroom, such as art, science, and everyday life.	GEO-4; GEO-6; GEO-7; GEO-13; GEO-16

*Reprinted with permission from *Principles and Standards for School Mathematics,* copyright 2000 by the National Council of Teachers of Mathematics. All rights reserved. Standards are listed with the permission of the National Council of Teachers of Mathematics (NCTM). NCTM does not endorse the content or validity of these alignments.

*NCTM Standards/Grades 6-8 Expectations	Lessons
Algebra	
Understand patterns, relations, and functions.	
Represent, analyze, and generalize a variety of patterns with tables, graphs, words, and, when possible, symbolic rules.	ALG-16
Relate and compare different forms of representation for a relationship.	ALG-10; ALG-11; ALG-12; ALG-13; ALG-14
Identify functions as linear or nonlinear and contrast their properties from tables, graphs, or equations.	ALG-16
Represent and analyze mathematical situations and structures using algebraic symbols.	
Develop an initial conceptual understanding of different uses of variable.	ALG-5
Explore relationships between symbolic expressions and graphs of lines, paying particular attention to the meaning of intercept and slope.	
Use symbolic algebra to represent situations and to solve problems, especially those that involve linear relationships.	ALG-5
Recognize and generate equivalent forms for simple algebraic expressions and solve linear equations.	
Use mathematical models to represent and understand quantitative relationships.	
Model and solve contextualized problems using various representations, such as graphs, tables, and equations.	ALG-5; ALG-15; ALG-16
Analyze change in various contexts.	
Use graphs to analyze the nature of changes in quantities in linear relationships.	ALG-16

*NCTM Standards/Grades 6-8 Expectations	Lessons
Measurement	
Understand measurable attributes of objects and the units, systems, and processes of measurement.	
Understand both metric and customary systems of measurement.	MEA-7; MEA-8; MEA-9
Understand relationships among units and convert from one unit to another within the same system.	
Understand, select, and use units of appropriate size and type to measure angles, perimeter, area, surface area, and volume.	MEA-5; MEA-6; MEA-7
Apply appropriate techniques, tools, and formulas to determine measurements.	
Use common benchmarks to select appropriate methods for estimating measurements.	
Select and apply techniques and tools to accurately find length, area, volume, and angle measures to appropriate levels of precision.	MEA-6; MEA-7; MEA-8; MEA-9

*Reprinted with permission from *Principles and Standards for School Mathematics,* copyright 2000 by the National Council of Teachers of Mathematics. All rights reserved. Standards are listed with the permission of the National Council of Teachers of Mathematics (NCTM). NCTM does not endorse the content or validity of these alignments.

Measurement Continued	
Develop and use formulas to determine the circumference of circles and the area of triangles, parallelograms, trapezoids, and circles and develop strategies to find the area of more-complex shapes.	MEA-8; MEA-9
Develop strategies to determine the surface area and volume of selected prisms, pyramids, and cylinders.	MEA-5; MEA-6; MEA-7
Solve problems involving scale factors, using ratio and proportion.	
Solve problems involving rates and derived measurements for such attributes as velocity and density.	

*NCTM Standards/Grades 6-8 Expectations	Lessons
Data Analysis and Probability	
Formulate questions that can be addressed with data and collect, organize, and display relevant data to answer them.	
Formulate questions, design studies, and collect data about a characteristic shared by two populations or different characteristics within one population.	
Select, create, and use appropriate graphical representations of data, including histograms, box plots, and scatterplots.	DAT-4
Select and use appropriate statistical methods to analyze data.	
Find, use, and interpret measures of center and spread, including mean and interquartile range.	DAT-1
Discuss and understand the correspondence between data sets and their graphical representations, especially histograms, stem-and-leaf plots, box plots, and scatterplots.	
Develop and evaluate inferences and predictions that are based on data.	
Use observations about differences between two or more samples to make conjectures about the populations from which the samples were taken.	
Make conjectures about possible relationships between two characteristics of a sample on the basis of scatterplots of the data and approximate lines of fit.	DAT-2
Use conjectures to formulate new questions and plan new studies to answer them.	
Understand and apply basic concepts of probability.	
Understand and use appropriate terminology to describe complementary and mutually exclusive events.	DAT-7
Use proportionality and a basic understanding of probability to make and test conjectures about the results of experiments and simulations.	DAT-6; DAT-8
Compute probabilities for simple compound events, using such methods as organized lists, tree diagrams, and area models.	DAT-5; DAT-8

*Reprinted with permission from *Principles and Standards for School Mathematics*, copyright 2000 by the National Council of Teachers of Mathematics. All rights reserved. Standards are listed with the permission of the National Council of Teachers of Mathematics (NCTM). NCTM does not endorse the content or validity of these alignments.

Number and Operations

The concept of **Number** pervades all areas of mathematics and is, therefore, a cornerstone of K-8 mathematics education. **Operations** – the use of numbers to add, subtract, multiply, divide, and perform operations with powers and roots, give students tools to solve real-life problems. Together, **Numbers and Operations** combine to form the core of mathematics instruction to give students greater number sense and more fluency in performing arithmetic operations.

Developing number sense is still a main goal of mathematics instruction in grades 5 and 6. Fractions in particular, pose challenges in all their applications – computation, comparison, order, and estimation. Manipulatives provide visual demonstrations that the same quantity can be represented by various equivalent number forms. In this way, students start to gain proficiency in translating among fractions, decimals, and percents, and learn to choose the form most appropriate and convenient for a practical situation.

> **The NCTM Standards for Number and Operations suggest that students should**
> - Understand numbers, ways of representing numbers, relationships among numbers, and number systems
> - Understand meanings of operations and how they relate to one another
> - Compute fluently and make reasonable estimates

At this level, students use tools such as estimation, mental computation, pencil-and-paper, manipulatives, and technology to solve problems. Students' understanding of fractions, decimals, and percents is challenged and expanded, while their experiences with ratio and proportion allow them to use proportionality to approach problems involving concepts in geometry, such as similarity. The following activities are built around manipulatives that students can use to develop skills and explore concepts in **Number and Operations**.

Number and Operations

Contents

Lesson 1 Reason with Fractions 20
Objective: Reason with fractions by comparing parts of a whole.
Manipulative: fraction circles

Lesson 2 Fractional Parts of a Collection 22
Objective: Use fractions to describe parts of a collection.
Manipulative: centimeter cubes

Lesson 3 Equivalent Fractions 24
Objective: Generate equivalent fractions.
Manipulative: fraction squares

Lesson 4 Decimals 26
Objective: Understand place value in decimal numbers.
Manipulative: base-ten blocks

Lesson 5 Equivalent Fractions and Decimals 28
Objective: Identify equivalent fractions and decimals.
Manipulative: fraction circles and Fraction Tower® Equivalency Cubes

Lesson 6 Compare and Order Fractions and Decimals 30
Objective: Compare and order fractions and decimals.
Manipulative: Fraction Tower® Equivalency Cubes

Lesson 7 Percents, Fractions, and Decimals 32
Objective: Write a number as a percent, a fraction, and a decimal.
Manipulative: fraction circles and fraction circle rings

Lesson 8 Mixed Numbers, Decimals, and Percents Greater than 100% 34
Objective: Write a number as a mixed number, a decimal, and a percent greater than 100%.
Manipulative: fraction squares

Lesson 9 Factors, Primes, and Prime Factorization 36
Objective: Determine if a number is prime or composite and express the prime factorization of a number.
Manipulative: color tiles

Lesson 10 Squares and Square Roots 38
Objective: Find the square of a number and the square root of a perfect square.
Manipulative: color tiles

Lesson 11 Add Fractions with Unlike Denominators 40
Objective: Add fractions with unlike denominators.
Manipulative: fraction circles

Lesson 12 Subtract Fractions with Unlike Denominators 42
Objective: Subtract fractions with unlike denominators.
Manipulative: fraction squares

Lesson 13 Add and Subtract Decimals 44
Objective: Add and subtract decimals involving tenths and hundredths.
Manipulatives: base-ten blocks

Lesson 14 Multiply with Fractions 46
Objective: Multiply with fractions, including a fraction and a whole number, and a fraction with a fraction.
Manipulative: Fraction Tower® Equivalency Cubes

Lesson 15 Meaning of Division 48
Objective: Explore the meaning of division.
Manipulative: centimeter cubes

Lesson 16 Divide with Fractions 50
Objective: Divide with fractions.
Manipulative: fraction squares

Lesson 17 Multiply and Divide Decimals 52
Objective: Multiply and divide decimals to hundredths.
Manipulative: base-ten blocks

Lesson 18 Ratios 54
Objective: Use ratios to represent relationships.
Manipulative: Cuisenaire® Rods

Lesson 19 Proportions 56
Objective: Use proportions to represent relationships.
Manipulative: Cuisenaire® Rods

Lesson 1

Number and Operations

Reason with Fractions

The understanding of a fraction as part of a whole is essential to all other work with fractions. By using various area models for parts of a whole, students see how fractions relate to the whole and to each other. Here, students are encouraged to think flexibly to compare fractions. They apply their understanding of fractions to reason about the amount that a fraction represents.

Objective
Reason with fractions by comparing parts of a whole.

Skills
- Representing rational numbers
- Comparing fractions
- Reasoning

NCTM Expectations
Grades 3–5
Number and Operations
- Develop understanding of fractions as parts of unit wholes, as parts of a collection, as locations on number lines, and as divisions of whole numbers.
- Use models, benchmarks, and equivalent forms to judge the size of fractions.

Try It! Perform the Try It! activity on the next page.

Talk About It
Discuss the Try It! activity.

- **Ask:** *Which is smaller, $\frac{1}{6}$ of a pie or $\frac{1}{8}$ of a pie?*
- **Ask:** *Which is greater, $\frac{5}{6}$ of a pie or $\frac{7}{8}$ of a pie?*
- **Ask:** *How are your answers to these two questions related?*

Solve It
Reread the problem with students. Students should reason that since the slice taken from the blueberry pie is smaller, more blueberry pie remains. Each pie has one fewer slice, but the removed slices were not the same size. Have students write the inequalities used to solve this problem: $\frac{1}{8} < \frac{1}{6}$ and $\frac{7}{8} > \frac{5}{6}$.

More Ideas
For other ways to teach about reasoning with fractions—

- Use fraction squares instead of fraction circles to solve this problem. Compare the rectangles that represent $\frac{1}{6}$ and $\frac{1}{8}$.

- Using Cuisenaire® Rods, tell students that each of the orange rods represents one whole. Ask students which rods represent $\frac{1}{2}$. Have students demonstrate their answers by laying two of these rods beneath an orange rod. Repeat this approach for $\frac{1}{5}$ and $\frac{1}{10}$. Ask students to compare the rods that represent $\frac{1}{2}$, $\frac{1}{5}$, and $\frac{1}{10}$.

Standardized Practice
Have students try the following problem.

Which shows the fractions listed in order from least to greatest?

A. $\frac{1}{5}, \frac{1}{6}, \frac{4}{5}, \frac{5}{6}$

B. $\frac{1}{6}, \frac{1}{5}, \frac{4}{5}, \frac{5}{6}$

C. $\frac{1}{6}, \frac{1}{5}, \frac{5}{6}, \frac{4}{5}$

D. $\frac{1}{6}, \frac{5}{6}, \frac{1}{5}, \frac{4}{5}$

Try It! 20 minutes | Groups of 4

Here is a problem about reasoning with fractions.

Abby made two pies of equal size. She cut the apple pie into six equal slices and the blueberry pie into eight equal slices. Then she ate one slice from each pie. Which one, apple or blueberry, has more pie remaining?

Introduce the problem. Then have students do the activity to solve the problem. Distribute fraction circles, paper, and pencils to students.

Materials
- fraction circles (1 set per group)
- paper (1 sheet per group)
- pencils (1 per group)

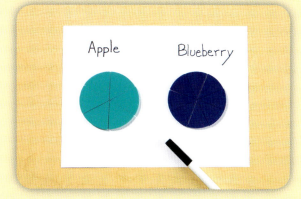

1. Say: *In the problem, the apple pie is cut into 6 equal slices. The blueberry pie is cut into 8 equal slices. Use a fraction circle to represent each pie.* Have students place their circles on a piece of paper and label them *Apple* and *Blueberry*.

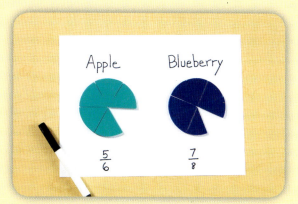

2. Say: *Next in the problem, Abby eats one slice from each pie. Show this with your fraction circles. Then, below each "pie", write a fraction to represent the remaining amount.* Students remove one slice from each circle and write two fractions.

⚠ Look Out!

Students may have difficulty deciding that $\frac{1}{8}$ is less than $\frac{1}{6}$, because they start with the knowledge that 8 is *greater* than 6. You can use fraction circles to show that as a circle is divided into more parts, the parts become smaller, so the fractions that represent the parts also get smaller. Students may easily understand that in order for a pie to have more slices, the slices must be smaller.

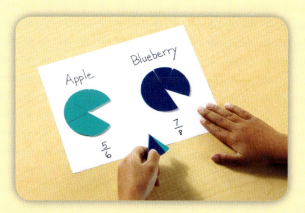

3. Say: *Now look at the slices you removed. Which slice is smaller, $\frac{1}{6}$ or $\frac{1}{8}$? Use this information to decide which pie has more remaining.* Students can place the fractional parts next to each other, or on top of each other, to decide which is smaller.

Lesson 2

Number and Operations

Fractional Parts of a Collection

Using a fraction to describe part of a collection can help students understand the part-whole relationship. A collection can be of any size, and a fraction can be used to describe any part of it. In this activity, students are given information about two parts of a set. They build the set by parts and must find the total. Students see that the fraction describing one part of the set plus the fraction describing the other part equals 1 (the whole set).

Try It! Perform the Try It! activity on the next page.

Objective
Use fractions to describe parts of a collection.

Skills
- Representing rational numbers
- Working with fractions
- Using patterns

NCTM Expectation
Grades 3–5
Number and Operations
- Develop understanding of fractions as parts of unit wholes, as parts of a collection, as locations on number lines, and as divisions of whole numbers.

Talk About It
Discuss the Try It! activity.

- **Ask:** *How many red cubes did you use for students wearing red shirts?*
- **Ask:** *How many cubes did you use for students not wearing red shirts? Does it matter which color cubes you used for these students?*
- **Ask:** *How many cubes represent the whole class? Do you need this number to solve the problem? Why or why not?*

Solve It
Reread the problem with students. Since there are 4 red cubes, and a total of 25 cubes, the fraction that describes the red cubes in the collection is $\frac{4}{25}$. So, the fraction of students in the class who are wearing red shirts is $\frac{4}{25}$. Note that the fraction of students in the class who are not wearing red shirts is $\frac{21}{25}$.

More Ideas
For other ways to teach fractional parts of a collection—

- Have students use centimeter cubes to create a collection that includes 2 red, 5 blue, 4 yellow, 1 green, 3 brown, 6 pink, 3 orange, 1 white, and 4 black cubes. Ask students to write a fraction for each color in the collection. Note that all the fractions added together equal 1.
- Have students use two-color counters, with red representing boys and yellow representing girls. Students can use the counters to show the number of boys and the number of girls in their math classroom and then write a fraction to describe each part.

Standardized Practice
Have students try the following problem.

There are 25 students in a class. If there are 16 girls in the class, what fraction of the class is girls?

A. $\frac{9}{41}$ B. $\frac{9}{25}$ C. $\frac{16}{41}$ D. $\frac{16}{25}$

Try It! 15 minutes | Pairs

Here is a problem about fractional parts of a collection.

Four students in a class are wearing red shirts. Twenty-one students in the class are not wearing red shirts. What is the fraction of students in the class wearing red shirts?

Introduce the problem. Then have students do the activity to solve the problem. Distribute about 40 centimeter cubes, including at least 4 red cubes and 21 non-red cubes, to each pair. Give each pair a sheet of paper and a pencil.

Materials
- centimeter cubes (40 per pair)
- paper (1 sheet per pair)
- pencils (1 per pair)

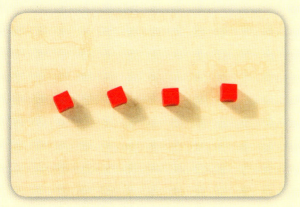

1. Say: *Use red cubes to represent the students who are wearing red shirts.* Have students place 4 red cubes together in a group.

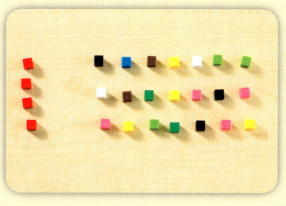

2. Say: *In a different area, use cubes to represent the students who are not wearing red shirts.* Have students place 21 non-red cubes in a different group.

3. Say: *Move all the red and non-red cubes together. Find the total number of cubes.* Students may move the cubes together in any way. Have them write the number of cubes in each group (Red, Not Red, and Total) and determine the answer to the problem.

⚠ Look Out!

Students might think that $\frac{4}{21}$, not $\frac{4}{25}$, is the correct answer to this problem. Show and explain the meaning of each of the following fractions.

$\frac{4}{25}$ ⟶ Red/Total

$\frac{21}{25}$ ⟶ Not Red/Total

$\frac{4}{21}$ ⟶ Red/Not Red

The fraction $\frac{4}{21}$ is the ratio of red cubes to non-red cubes. To use a fraction to describe part of a set, you must compare the relevant part (red, in this case) to the total.

Number and Operations

Lesson 3

Number and Operations

Equivalent Fractions

Students need to be able to recognize, generate, and use equivalent fractions. Equivalent fractions are used in comparing and ordering; simplifying; and adding, subtracting, multiplying, and dividing fractions. This activity will prepare students to see and use equivalent fractions as equal ratios that form a proportion.

Objective
Generate equivalent fractions.

Skills
- Representing rational numbers
- Working with fractions
- Using patterns

NCTM Expectations

Grades 3–5
Number and Operations
- Recognize equivalent representations for the same number and generate them by decomposing and composing numbers.
- Use models, benchmarks, and equivalent forms to judge the size of fractions.

Try It! Perform the Try It! activity on the next page.

Talk About It
Discuss the Try It! activity.

- **Ask:** *If Bonnie has 2 children, how much land will each child get? How much land will the 2 children get together?*
- **Ask:** *If Bonnie has 3 children, how much land will each child get? How much land will the 3 children get together?*
- **Say:** *Continue and describe the patterns in this problem.*

Solve It

Reread the problem with students. In every case, the children will get half of the grandfather's land. As that half is divided among more and more children, the half is divided into more and more pieces. By solving this problem, students generate the following equivalent fractions:
$\frac{1}{2} = \frac{2}{4} = \frac{3}{6} = \frac{4}{8} = \frac{5}{10} = \frac{6}{12}$.

More Ideas

For other ways to teach equivalent fractions—

- Have students do this problem again using fraction circles. They can stack the equal areas for $\frac{1}{2}$ and spread the colors out to see the increase in the number of pieces for 2, 3, 4, 5, and 6 children. Ask students what the fraction would be for 10 children.

- Use fraction circles and fraction squares to generate other equivalent fractions, such as: $\frac{1}{3} = \frac{2}{6} = \frac{4}{12}$, $\frac{2}{3} = \frac{4}{6} = \frac{8}{12}$, $\frac{1}{4} = \frac{2}{8} = \frac{3}{12}$, $\frac{3}{4} = \frac{6}{8} = \frac{9}{12}$, $\frac{1}{5} = \frac{2}{10}$, $\frac{2}{5} = \frac{4}{10}$, $\frac{3}{5} = \frac{6}{10}$, $\frac{4}{5} = \frac{8}{10}$, $\frac{1}{6} = \frac{2}{12}$, $\frac{5}{6} = \frac{10}{12}$.

Standardized Practice

Have students try the following problem.

Which fraction is equivalent to $\frac{3}{4}$?

A. $\frac{6}{9}$ B. $\frac{9}{12}$ C. $\frac{8}{10}$ D. $\frac{5}{6}$

Try It! 15 minutes | Groups of 5

Here is a problem about equivalent fractions.

Bonnie bought half of her father's land and will divide it evenly among her own children. What fraction describes the portion of their grandfather's land that each child will receive if Bonnie has 2 children? 3 children? 4 children? 5 children? 6 children?

Introduce the problem. Then have students do the activity to solve the problem. Each student in the group can answer one question. Distribute the fraction squares, paper, and pencils to students.

Materials
- fraction squares (1 set per group)
- paper (2 sheets per group)
- pencils (1 per group)

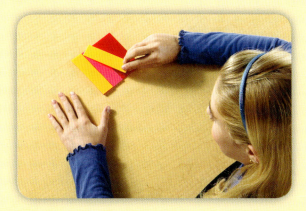

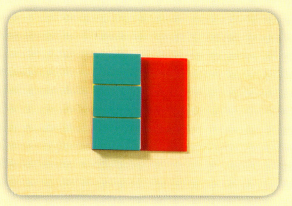

1. Say: *Use the red square to represent Bonnie's father's land. Place a pink rectangle on the square to represent Bonnie's land, which is half of her fathers land. Now show how Bonnie's land can be divided evenly for 2 children.* Students place 2 yellow rectangles on top of the pink rectangle to show $\frac{1}{2} = \frac{2}{4}$.

2. Ask: *How can you show that Bonnie's land is divided into 3 equal parts for 3 children?* Students place 3 aqua rectangles on top of the 2 yellow rectangles to show $\frac{1}{2} = \frac{2}{4} = \frac{3}{6}$.

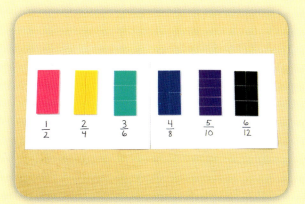

3. Say: *Continue the problem for 4, 5, and 6 children.* Students continue with blue, purple, and then black rectangles. Have students undo the stack, line up the colors on paper, and write a fraction for each color.

⚠ Look Out!

Students might think that 2 children will each get $\frac{1}{2}$ of the grandfather's land, but they will each get $\frac{1}{2}$ of half the grandfather's land, which is $\frac{1}{4}$. Together, the 2 children get $\frac{2}{4}$, or $\frac{1}{2}$, of the land. Similarly, 3 children will each get $\frac{1}{3}$ of half the land, or $\frac{1}{6}$. Together, the 3 children get $\frac{3}{6}$, or $\frac{1}{2}$, of the land. It follows that 4 children will each get $\frac{1}{4}$ of half, or $\frac{1}{8}$. Together, the 4 children get $\frac{4}{8}$, or $\frac{1}{2}$, of the land. Help students see and understand this pattern.

Lesson 4

Number and Operations

Decimals

Students need to understand decimal place value in order to work effectively with decimals. Decimal place value is simply an extension of whole-number place value, which students have studied previously. Common models used to help students understand place value are base ten blocks, which work with both decimals and whole numbers. In this activity, students use the blocks to model tenths, hundredths, and thousandths.

Try It! Perform the Try It! activity on the next page.

Objective

Understand place value in decimal numbers.

Skills

- Representing decimals
- Modeling place value
- Using multiple representations

NCTM Expectation

Grades 3–5
Number and Operations
- Understand the place-value structure of the base-ten number system and be able to represent and compare whole numbers and decimals.

Talk About It

Discuss the Try It! activity.

- **Ask:** *How many tenths (flats) are there in one whole (cube)?*
- **Ask:** *How many hundredths (rods) are there in one whole (cube)?*
- **Ask:** *How many thousandths (units) are there in one whole (cube)?*
- Tell students that the expression 0.1 + 0.02 + 0.005 is an expanded form of the decimal number 0.125. They may recall writing whole numbers in expanded form, such as 419 = 400 + 10 + 9.

Solve It

Reread the problem with students. Note that two whole numbers are given in this problem, 1000 and 8. By dividing 1000 by 8, you get another whole number, 125. To represent this as a part of the whole, you can write the fraction $\frac{125}{1000}$ and the decimal 0.125. Both numbers mean "one hundred twenty-five thousandths." Students break 0.125 into place-value parts to model this number using 125 total units.

More Ideas

For other ways to teach decimals—

- Have students use base ten blocks to model other decimal numbers, such as 0.6, 0.42, 0.07, 0.305, 0.009, and 0.231.
- Have students use base ten blocks to show that 1 tenth (flat) = 10 hundredths (rods) = 100 thousandths (units). Then have them write the corresponding decimal numbers, 0.1 = 0.10 = 0.100, and the corresponding fractions, $\frac{1}{10} = \frac{10}{100} = \frac{100}{1000}$.

Standardized Practice

Have students try the following problem.

Which of the following is an expanded form of 0.408?

A. 0.4 + 0.8   **C.** 0.04 + 0.008

B. 0.4 + 0.08   **D.** 0.4 + 0.008

Try It! 20 minutes | Groups of 3

Here is a problem about decimals.

A race is limited to 1,000 cyclists divided into 8 equal groups. Each group will start the race at a different time. The first group starts at 8:00 AM, the second group starts at 8:15 AM, the third group starts at 8:30 AM, and so on. Write and model a decimal number for the fraction of the 1,000 cyclists that is represented by each group.

Introduce the problem. Then have students do the activity to solve the problem. Distribute base ten blocks, worksheets, paper, and pencils. Display the large cube and say it represents the whole (1,000 cyclists). Have students determine the values of the other blocks.

Materials
- base ten blocks (1 flat, 10 rods, and 20 units per group, and 1 large cube for teacher demonstration)
- Decimal Models worksheet (BLM 1, 1 per group)
- paper (1 sheet per group)
- colored pencils (1 per group)

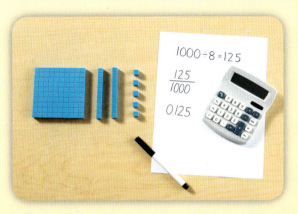

1. Say: *Calculate the number of cyclists in each group.* Students divide 1,000 by 8 and get 125. **Ask:** *If there are 125 cyclists in each group and 1,000 cyclists altogether, what fraction does each group represent? Write the fraction.* Students write $\frac{125}{1000}$.

2. Say: *Model this fraction as a decimal. The decimal has three parts: tenths, hundredths, and thousandths.* Students use 1 flat, 2 rods, and 5 units. **Say:** *Write the decimal.*

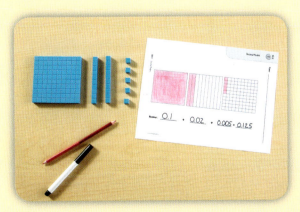

⚠ Look Out!

Students might think that the 1 flat, 2 rods, and 5 units represent the whole number 125, rather than the decimal number 0.125. This is good mathematical thinking, because these blocks can represent both 125 and 0.125! Explain this to the students in terms of place value.

3. Say: *Now, color and complete the worksheet to match the blocks you used to represent 0.125.* Students complete the worksheet.

Lesson 5

Number and Operations

Equivalent Fractions and Decimals

In this activity, students explore the relationship between fractions and decimals. They see that a fraction is equivalent to another fraction whose decimal equivalent is easy to write. Students then need to understand the meaning of fraction as a quotient of two numbers, using division as a way to get the equivalent decimal.

Try It! Perform the Try It! activity on the next page.

Objective

Identify equivalent fractions and decimals.

Skills

- Writing equivalent forms
- Representing rational numbers
- Using place value

NCTM Expectations

Grades 3–5
Number and Operations
- Develop understanding of fractions as parts of unit wholes, as parts of a collection, as locations on number lines, and as divisions of whole numbers.
- Use models, benchmarks, and equivalent forms to judge the size of fractions.
- Recognize and generate equivalent forms of commonly used fractions, decimals, and percents.

Talk About It

Discuss the Try It! activity.

- **Ask:** *How do the fraction circles help you solve this problem? How do the Fraction Tower® Equivalency Cubes help?*
- **Ask:** *How can you solve this problem without using fraction circles or Fraction Tower® Equivalency Cubes?* Students can solve the problem using division; $2 \div 5$.
- **Say:** *Charlie's friend Ben ate 3 out of 5 pieces of the quesadilla. What portion of the quesadilla did Ben eat? Write this as a fraction and as a decimal.*

Solve It

Reread the problem with students. Charlie ate $\frac{2}{5}$ of the quesadilla, which is a fraction in simplest form. By finding the equivalent fraction $\frac{4}{10}$ and saying "four tenths", students are led to write the equivalent decimal 0.4. Use the equivalency cubes to have students verify their answer.

More Ideas

For other ways to teach equivalent fractions and decimals—

- List the fractions $\frac{3}{4}, \frac{4}{5}, \frac{7}{10},$ and $\frac{3}{8}$. Have students use fraction squares placed on a base ten flat to find equivalent decimals for the fractions. Students may need guidance with $\frac{3}{8}$. Have students consider $\frac{2}{3}, \frac{5}{6},$ and $\frac{11}{12}$ using division. Help students identify the terminating and the repeating decimals.
- Students can use Fraction Tower Equivalency Cubes to help them memorize the commonly used fraction and decimal equivalents.

Standardized Practice

Have students try the following problem.

Amy bought $\frac{3}{8}$ yard of ribbon. The receipt shows an equivalent decimal for $\frac{3}{8}$. Which decimal is on the receipt?

A. 0.38   **B.** 0.375   **C.** 0.3   **D.** 0.266

28

Try It!   15 minutes | Groups of 4

Here is a problem about equivalent fractions and decimals.

Charlie cut a quesadilla into 5 pieces and ate 2 of them. What portion of the quesadilla did charlie eat? Write your answer as a fraction and a decimal.

Introduce the problem. Then have students do the activity to solve the problem. Distribute fraction circles, Fraction Tower® Equivalency Cubes, paper, and pencils to students.

Materials
- fraction circles (1 set per group)
- Fraction Tower® Equivalency Cubes (1 set per group)
- paper (1 sheet per group)
- pencils (1 per group)

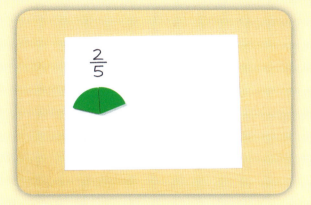

1. Say: *Charlie ate 2 pieces of the quesadilla. Write a fraction for the portion of the quesadilla that Charlie ate. Model this fraction with fraction circles.* Students write $\frac{2}{5}$ and select two parts (fifths) from the green fraction circles.

2. Say: *Decimal numbers are based on tens. Use the fraction circle divided into ten parts to model the amount of the quesadilla that Charlie ate. Write the fraction.* Students select four parts (tenths) from the purple fraction circle and write $\frac{4}{10}$. **Ask:** *How many tenths are there? Write the decimal.* Students say "four tenths" and write 0.4.

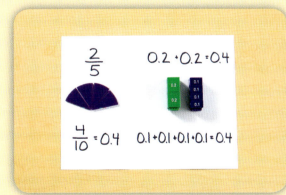

3. Say: *Now use two colors of Fraction Towers to model the amount of the quesadilla that Charlie ate. They must have equal heights.* Students use two green fifths and four purple tenths. **Say:** *Show the decimal numbers on your towers. Add the decimals for each color.* Students write 0.2 + 0.2 = 0.4 for green and 0.1 + 0.1 + 0.1 + 0.1 = 0.4 for purple.

⚠ Look Out!

Students often have a difficult time dividing a smaller number by a larger number. The fraction $\frac{2}{5}$ is the same as the division expression 2 ÷ 5, not 5 ÷ 2. When students try to use long division to calculate the decimal for $\frac{2}{5}$, they might write $2\overline{)5}$ rather than $5\overline{)2}$. Remind students to write the decimal point and zeros, then divide:

$$5\overline{)2.0}^{\,.4}$$

Number and Operations

Lesson 6

Number and Operations

Compare and Order Fractions and Decimals

The number line is a useful tool for comparing and ordering fractions and decimals. Students should be able to draw and mark a number line with fractions, with decimals, and with a combination of both types of numbers. They should be able to translate between forms and use whichever form is more convenient or more appropriate in the context of the problem.

Try It! *Perform the Try It! activity on the next page.*

Objective
Compare and order fractions and decimals.

Skills
- Comparing and ordering numbers
- Writing equivalent forms
- Graphing on a number line

NCTM Expectations
Grades 6–8
Number and Operations
- Work flexibly with fractions, decimals, and percents to solve problems.
- Compare and order fractions, decimals, and percents efficiently and find their approximate locations on a number line.

Talk About It
Discuss the Try It! activity.

- **Say:** *One pink Tower piece represents $\frac{1}{2}$. What decimal is equivalent to $\frac{1}{2}$?*
- **Say:** *Three green Tower pieces together represent $\frac{3}{5}$. What decimal is equivalent to $\frac{3}{5}$?* Note that each green Tower piece is $\frac{1}{5}$, or 0.2. Three pieces equal $\frac{1}{5} + \frac{1}{5} + \frac{1}{5} = \frac{3}{5}$, or 0.2 + 0.2 + 0.2 = 0.6.

Solve It
Reread the problem with students. Students find the equivalent decimals for the fractions using the Equivalency Cubes. Students locate these decimals on a number line, reading from left to right: 0.375, 0.5, 0.6, 0.75, 0.8. Next they write the foods in order: salami, Swiss, cheddar, turkey, ham.

More Ideas
For other ways to teach about comparing and ordering fractions and decimals—

- Have students use base ten blocks to compare fractions and decimals. Provide students with the decimals as fractions in simplest form. For example, note that the fractions are difficult to order. Next, have students write each number as a fraction with denominator 100 and model with base ten blocks $\frac{1}{2}, \frac{3}{4}, \frac{2}{5}$. These fractions are easier to order.

- Have students use the fraction and decimal fraction circle rings to compare the sizes of the fraction circle pieces in the set. Ask students to place each piece inside the fraction circle ring and mark and label each piece starting from the 0. Then have students use the decimal circle ring to label each of the fraction pieces they marked.

Standardized Practice
Have students try the following problem.

Which of the following statements is true?

A. $0.5 < \frac{5}{8}$ B. $0.15 > \frac{3}{10}$ C. $\frac{4}{5} > 0.8$ D. $\frac{1}{3} < 0.25$

Try It! 15 minutes | Pairs

Here is a problem about ordering fractions and decimals.

Bob is making a snack tray for a party. He bought packaged cheese; $\frac{1}{2}$ pound of Swiss and $\frac{3}{5}$ pound of cheddar. The deli clerk sliced meats for Bob. Bob got 0.8 pound of ham and 0.75 pound of turkey, and 0.375 pound of salami. Write the foods in order from least to greatest weight.

Introduce the problem. Then have students do the activity to solve the problem. Give each pair of students Fraction Tower® Equivalency Cubes, a straightedge, paper, and pencils.

Materials
- Fraction Tower® Equivalency Cubes (1 set per pair)
- straightedges (1 per pair)
- paper (1 sheet per pair)
- pencils (1 per pair)

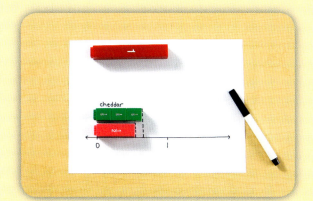

1. Say: *Select Fraction Towers to represent the Swiss cheese and the cheddar cheese. Draw a blank number line. Mark zero. Mark 1 using a red Tower. Use the Towers to draw a segment for each type of cheese.* Students trace along the tops of the pink and green Towers and write *Swiss* and *cheddar*.

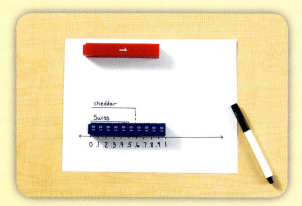

2. Say: *Next, use the purple Fraction Tower to mark all the tenths on your number line between 0 and 1.* Students mark and label 0.1, 0.2, 0.3, 0.4, 0.5, 0.6, 0.7, 0.8, 0.9, and 1.

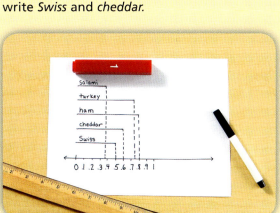

3. Say: *Draw and label a segment for each meat. Notice that 0.75 falls between two numbers and that 0.375 does too. Write the foods in order from least to greatest weight.*

⚠ Look Out!

Make sure students think about place value when they compare decimals such as 0.8 (ham) and 0.375 (salami). Some students might think that 0.8 < 0.375 because 8 < 375. Point out that 0.8 = 0.800, so they should compare 0.800 with 0.375. Since 800 > 375, then 0.800 > 0.375, and therefore 0.8 > 0.375. When comparing 0.75 (ham) with 0.8 (cheddar), students should see that there's a little more ham than cheddar. In this case, 0.80 > 0.75 because 80 > 75.

Lesson 7

Number and Operations

Percents, Fractions, and Decimals

Percents, fractions, and decimals are different ways to show the same value. Understanding these different ways to represent numbers will prepare students to work flexibly with numbers to describe and compare data, name probabilities, and utilize the concept of proportionality. It is important to establish the relationship between these forms to prepare students to study percents less than 1% and greater than 100%.

Try It! *Perform the Try It! activity on the next page.*

Objective
Write a number as a percent, a fraction, and a decimal.

Skills
- Representing numbers
- Comparing numbers
- Converting fractions and decimals

NCTM Expectation
Grades 3–5
Number and Operations
- Recognize and generate equivalent forms of commonly used fractions, decimals, and percents.

Talk About It
Discuss the Try It! activity.

- **Ask:** *How do you write $\frac{1}{5}$ as a decimal and as a percent?*
- **Ask:** *If 55% of a circle is shaded, what percent is not shaded?*

Solve It
Reread the problem with students. Discuss how to find and represent the number of students who chose *Other Sport* in the survey. Note that students can subtract the sum of their shaded parts from 100%, or find fraction pieces to fill in the unshaded portion. Ask them to write an explanation of how they determined the order.

More Ideas
For other ways to teach equivalency among fractions, decimals, and percents—

- Have students use Fraction Tower® Equivalency Cubes to model the problem. Begin by having students connect pieces for 20%, $\frac{1}{10}$, and 0.25. Compare the tower to one whole to find how they can represent the students who selected the category *Other Sport*. Have students write the equation shown by the tower. Then have students combine pieces to find equivalents for each survey response. Discuss how to write numbers in different forms.

- Have students cover base-ten flats with fraction squares to represent decimal numbers. Have students write the decimal and fraction for each example. Introduce the term *percent*. Elicit that the percent is the number of covered squares on the flat, or the number of hundredths that are covered.

Standardized Practice
Have students try the following problem.

Which percent is equal to $\frac{3}{25}$?

A. 3% B. 6% C. 12% D. 325%

Try It! 15 minutes | Groups of 4

Here is a problem about percents, fractions, and decimals.

In a survey about favorite sports, 20% of the students chose basketball, $\frac{1}{10}$ chose hockey, 0.25 chose soccer, and the rest chose the category "Other Sport." List the four responses from most popular to least popular sport.

Introduce the problem. Then have students do the activity to solve the problem. Distribute the fraction circles, fraction circle rings, worksheets, and colored pencils to students. Point out the symbol for the percent on the percent ring.

Materials
- fraction circles (1 set per group)
- fraction circle rings (1 set per group)
- Fraction Circle in 100ths worksheet (BLM 2, 1 per group)
- colored pencils (3 per group)

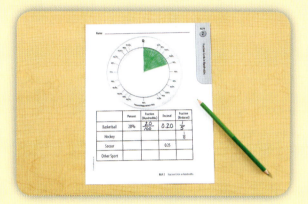

1. Have students use the percent ring as a guide to shade 20% of a hundredths circle.
Ask: *How many hundredths have you shaded?*
Say: *Write a fraction to represent your answer and then write it as a decimal.* Note the connection between percent, hundredths, and the position of the decimal point.
Say: *Write the fraction in reduced form.*

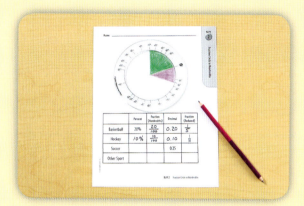

2. Have students use the fraction ring as a guide to shade $\frac{1}{10}$ of the hundredths circle (adjacent to the 20% that is already shaded).
Ask: *How many hundredths have you shaded?*
Say: *Write a fraction to represent your answer and write a decimal and a percent for this fraction.*

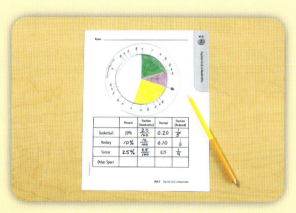

3. Have students use the decimal ring as a guide to shade 0.25 of the hundredths circle.
Ask: *How many hundredths have you shaded?*
Say: *Write a fraction to represent your answer and write a percent to represent this fraction. Then write the fraction in reduced form.*

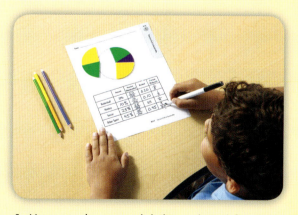

4. Have students model the reduced fractions by finding pieces from the appropriate fraction circles and matching them to the three shaded sections of the hundredths circle. **Ask:** *How do you determine the percent who chose "Other Sport?"* Have students complete the table.

Number and Operations

Lesson 8

Number and Operations

Mixed Numbers, Decimals, and Percents Greater than 100%

Students expand their experiences with different number representations by looking at mixed numbers and their equivalent decimals and percents. Using models helps students increase their flexibility with these numbers. As their number fluency increases, students begin to differentiate between situations in which one representation may be more suitable than another.

Objective
Write a number as a mixed number, a decimal, and a percent greater than 100%.

Try It! Perform the Try It! activity on the next page.

Skills
- Representing numbers
- Comparing numbers
- Decomposing numbers

NCTM Expectations
Grades 6–8
Number and Operations
- Work flexibly with fractions, decimals, and percents to solve problems.
- Compare and order fractions, decimals, and percents efficiently and find their approximate locations on a number line.
- Develop meaning for percents greater than 100 and less than 1.

Talk About It
Discuss the Try It! activity.

- **Ask:** *Could you compare the numbers in the same way if each person had used a different-sized carton to hold the balls? Why or why not?*
- Have students explain how to write $1\frac{1}{12}$ as a percent.
- Have students describe situations in which it might be better to choose one form over another to represent a number. For example, when using money, a decimal is the more accepted number representation.

Solve It
Reread the problem with students. Have students compare the fractional portions of their models to answer the question.

More Ideas
For other ways to teach equivalency among mixed numbers, decimals, and percents greater than 100%—

- Have students use Fraction Tower® Equivalency Cubes to model the problem. Have students see that one whole equals 1, 1.0, and 100%, regardless of the number of equal-size parts in the whole. Have students show amounts greater than one whole as they explore how different values can be combined to show mixed numbers, decimals greater than one, and percents greater than 100%.
- Have students cover base ten flats with fraction squares to model amounts greater than one whole, then write each amount in the three forms using their knowledge of percents, decimals, and fractions. Some students may find it helpful to shade 10 × 10 grids (BLM 3) to show each amount and to help them to rewrite the amounts in each of the three forms.

Standardized Practice
Have students try the following problem.

Which equation is true?

A. $1\frac{7}{10} = 1710\%$ B. $20.0 = 200\%$ C. $1\frac{1}{5} = 120\%$ D. $2.45 = 2.45\%$

Try It! 20 minutes | Groups of 4

Here is a problem about mixed numbers, decimals, and percents.

Randi, Nick, and LaKeisha collected golf balls from a pond on the golf course and put them into cartons. Each carton holds the same number of balls. Randi filled $1\frac{3}{8}$ cartons, Nick filled 1.5 cartons, and LaKeisha filled 175% of a carton. Who found the most golf balls?

Introduce the problem. Then have students do the activity to solve the problem. Distribute fraction squares, paper and pencils to students. Review the ways that one whole can be represented: 1, 1.0, and 100%. In the problem, whole is one carton of golf balls.

Materials
- fraction squares (1 set per group)
- paper (3 sheets per group)
- pencils (1 per group)

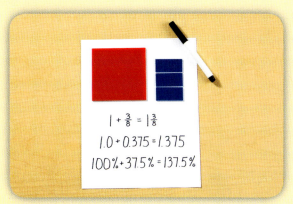

1. Write $1 + \frac{3}{8} = 1\frac{3}{8}$ on the board. **Say:** *A mixed number is the sum of its whole-number part and its fraction part. Use fraction pieces to show $1\frac{3}{8}$.* Have students rename the fraction part as a decimal and as a percent, use the correct names for the whole, and write the corresponding equations.

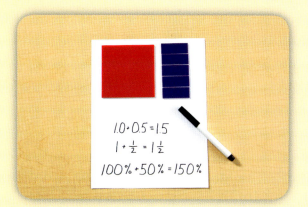

2. Have students show 1.5. **Ask:** *How can we write an equation to add the whole-number part and the decimal part of this number?* Write *1.0 + 0.5 = 1.5* on the board. **Ask:** *How can we write five-tenths as a fraction in simplest form and as a percent?* Have students write the three equations shown by the model.

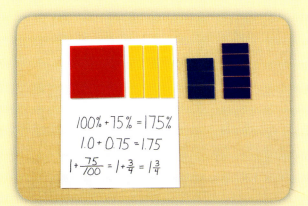

3. Write *100% + ? = 175%* on the board. **Ask:** *What percent can replace the question mark?* Discuss how to use fraction pieces to show 75%. Have students model the equation and write three equations shown by the model.

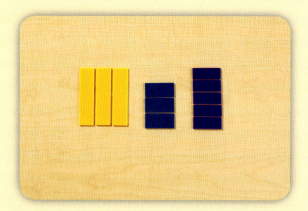

4. Have students recognize that the red square represents 1 and is the same in each case. Students can therefore answer the question by comparing the fractional portions of their models.

Lesson 9

Number and Operations

Factors, Primes, and Prime Factorization

Students learning about factors discover that some numbers have fewer factors than other numbers. A prime number has exactly two factors: the number 1 and the number itself. A composite number has more than two factors. Since the number 1 has only itself as a factor, it is neither prime nor composite. A number expressed as the product of prime factors is the prime factorization of that number.

Try It! Perform the Try It! activity on the next page.

Objective
Determine if a number is prime or composite and express the prime factorization of a number.

Skills
- Identifying factors
- Representing numbers
- Multiplying and dividing

NCTM Expectations
Grades 6–8
Number and Operations
- Use factors, multiples, prime factorization, and relatively prime numbers to solve problems.

Talk About It
Discuss the Try It! activity.

- **Ask:** *Why is 1 neither a prime number nor a composite number?*
- **Ask:** *How many even numbers are prime? How do you know?*
- **Ask:** *Why are the factors in the prime factorization written in order from least to greatest?*

Solve It
Reread the problem with students. Discuss how to write the prime factorization of a number. Be sure students realize that they can find the prime factorization using any of the arrays that they build. If necessary have them repeat the factorization starting with a different array.

More Ideas
For other ways to teach about factors, primes, and prime factorization—

- Have students use Snap Cubes® to make rectangles and record the side lengths. Ask students to compare the side lengths for the numbers that can make only one rectangle. Then introduce the terms *prime number* and *composite number*. Show students how to make a factor tree.

- Provide students with a Hundred Chart worksheet (BLM 5). Have students use divisibility rules to cross out numbers that are divisible by 2, 3, 4, and so on. Then have them use centimeter cubes to model the remaining numbers on this *Sieve of Eratosthenes* to confirm that they are prime numbers.

Standardized Practice
Have students try the following problem.

Which shows the prime factorization of 30?

A. 6×5　　B. 2×15　　C. $2 \times 3 \times 5$　　D. $1 \times 3 \times 10$

Try It! 30 minutes | Groups of 4

Here is a problem about factors, primes, and prime factorization.

Danielle wants to find all the prime numbers less than 20. She also wants to know how 24 can be written as a product of prime factors. How can she do this?

Introduce the problem. Then have students do the activity to solve the problem. Distribute color tiles, recording sheets, paper, and pencils to students. Review the term *factor* as a number that is multiplied to find a product.

Materials
- color tiles (100 per group)
- Tile Arrays recording sheet (BLM 4, 1 per group)
- paper (11" x 17" 1 sheet per group)
- pencils (1 per group)

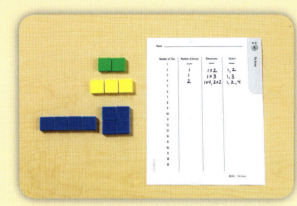

1. Have students make as many arrays as they can for each listed number of tiles and record their data. **Ask:** *For which numbers can you make only one array?* Introduce the terms *prime* and *composite*. Explain that 3 is prime since only one array can be made and 4 is composite since more than one array can be made.

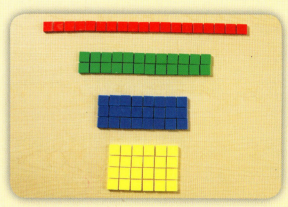

2. Tell students that some composite numbers, such as 24, have several possible arrays. Introduce the term *prime factorization*. **Say:** *An array for a composite number can be used to find the prime factorization of that number.* Have students make four arrays representing the number 24.

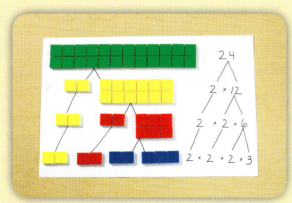

3. Say: *Transfer your 2 × 12 array to a sheet of paper.* Have students recognize that the side lengths are factors of 24 and have them build arrays for these factors. Have students repeat this reasoning for the two factors, and guide them to build a factor tree using tiles.

⚠ Look Out!

Students may include composite numbers in a prime factorization. Suggest that they look at the Sieve of Eratosthenes to help them verify that all the factors are prime numbers. Show students who forget to include a factor in the prime factorization how to make a factor tree to organize their work, and encourage them to multiply the factors listed to make sure that the product is correct.

Lesson 10

Number and Operations

Squares and Square Roots

Building squares helps students identify the relationship between squares and square roots. Students who can find the square roots of perfect squares are better prepared to estimate the square roots of other numbers. Learning these skills enables students to understand and use the Pythagorean relationship, the quadratic formula, and the distance formula.

Try It! *Perform the Try It! activity on the next page.*

Talk About It

Discuss the Try It! activity.

- **Ask:** *Why is a number written with an exponent of 2 often read as that number squared?*
- **Ask:** *What is the relationship between a square and its square root?*
- **Ask:** *How can you find the square root of a number if you do not have tiles to make a square?*

Solve It

Reread the problem with students. Discuss how the questions relate to finding squares and square roots. Have students write a paragraph using the terms *square* and *square root* to answer the Try It! questions.

More Ideas

For other ways to teach the relationship between squares and square roots—

- Have students use centimeter cubes to make square arrays and record their observations in a table. One row of the table will be side length, or Square Root, and the second row will be the total number of cubes, or Square. Encourage pairs to combine their cubes to make squares for larger numbers.
- Use Cuisenaire® Rods to have students model squares with side lengths up to 10 units. Have students group rods of equal length to make each square. The number of rods used equals the side length, or square root.
- Use geoboards to make four of the square designs in the problem. Instruct students to consider the space between pegs as one unit when making squares.

Standardized Practice

Have students try the following problem.

What is the square root of 100?

A. 1   **B.** 10   **C.** 1,000   **D.** 10,000

Objective

Find the square of a number and the square root of a perfect square.

Skills

- Representing numbers
- Using inverse relationships
- Using symbols to show number relationships

NCTM Expectation

Grades 6–8
Number and Operations
- Understand and use the inverse relationships of addition and subtraction, multiplication and division, and squaring and finding square roots to simplify computations and solve problems.

Try It! 25 minutes | Groups of 4

Here is a problem about squares and square roots.

Kayla uses 91 square tiles to make 6 designs. Each design is a square and all are different sizes. How many tiles are used in each square? What is the length of the sides in each square?

Introduce the problem. Then have students do the activity to solve the problem. Define the terms *square* and *square root*. Ask students to label one side of their papers Square and the other side Square Root. Distribute color tiles, grid paper, paper, and pencils to students.

Materials
- color tiles (91 per group)
- Inch Grid Paper (BLM 6, 1 sheet per group)
- paper (1 sheet per group)
- pencils (1 per group)
- colored pencils (2 per group)

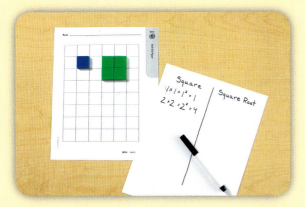

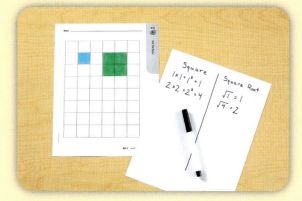

1. On the grid paper, have students build a 1×1 and a 2×2 square. **Ask:** *How can you find the number of tiles in each square?* Write $1 \times 1 = 1^2 = ?$ and $2 \times 2 = 2^2 = ?$ on the board. Review the meaning of the exponent 2. Have students write equations to show each square number.

2. Ask: *Which part of the square represents the square root?* Write $\sqrt{4} = 2$ on the board. Explain that the radical sign means find the square root. Have students copy the equation and write an equation showing the square root of 1 on their papers.

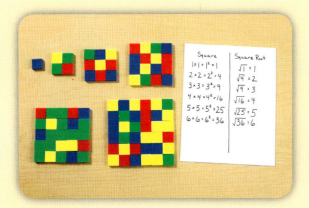

3. Have students use tiles to make the other squares. Remind them to include a one-tile square in their list and to use the appropriate symbols when writing equations for the square and the square root.

⚠ Look Out!

If students have difficulty finding the 6 smallest perfect squares, suggest that they start with one tile, add one tile to the row, and then add a tile to each column to make a square. Encourage them to continue this method to build larger squares. If students confuse the square and the square root, point out that in mathematics "root" means answer, so for a given number of tiles, the square root of that number gives the dimensions of the square that can be built.

Lesson 11

Number and Operations

Add Fractions with Unlike Denominators

Students build on their knowledge of fractions as they use models to add fractions with unlike denominators. They may use different approaches, such as number sense or reasoning, to find the solution to a problem. In this activity students follow the standard algorithm using their knowledge of equivalent fractions.

Try It! Perform the Try It! activity on the next page.

Objective
Add fractions with unlike denominators.

Skills
- Adding rational numbers
- Representing rational numbers
- Finding equivalent fractions

NCTM Expectations
Grades 3–5
Number and Operations
- Recognize and generate equivalent forms of commonly used fractions, decimals, and percents.
- Use visual models, benchmarks, and equivalent forms to add and subtract commonly used fractions and decimals.

Talk About It
Discuss the Try It! activity.
- **Ask:** *Why do we need common denominators?*
- Have students explain how they know which fractions to rename when they find the sum.
- **Ask:** *How do you use your knowledge of equivalent fractions to add fractions with unlike denominators?*

Solve It
Reread the problem with students. Have students draw a map of Emilio's ride and label the distances between destinations. Then have them explain how they found the total distance Emilio rode to school that morning.

More Ideas
For other ways to teach adding fractions—
- Use Fraction Tower® Equivalency Cubes to model the problems.
- Use the fraction ring from a set of fraction circle rings along with the fraction circles to model the problem. Students can place fraction pieces showing each addend inside the ring to find the sum. Make sure students understand why the denominator in the sum differs from the denominator in one or both of the addends.

Standardized Practice
Have students try the following problem.

Deon grows carrots in $\frac{1}{6}$ of his garden. He grows potatoes in another $\frac{1}{4}$ of the garden. The rest of the garden is planted with flowers. What fraction of Deon's garden is used to grow vegetables?

A. $\frac{1}{10}$ B. $\frac{2}{12}$ C. $\frac{2}{10}$ D. $\frac{5}{12}$

Try It! 15 minutes | Pairs

Here is a problem about adding fractions with unlike denominators.

Emilio rides $\frac{1}{4}$ mile from his house to his friend Jake's house. Together they ride $\frac{3}{8}$ mile to school. How far does Emilio ride to school that day?

Introduce the problem. Then have students do the activity to solve the problem. Distribute fraction circles, paper, and pencils to students.

Materials
- fraction circles (1 set per pair)
- paper (1 sheet per pair)
- pencils (1 per pair)

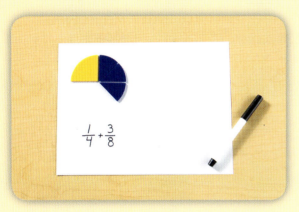

1. Ask: *How do you know what operation to use to solve the problem?* Write $\frac{1}{4} + \frac{3}{8}$ on the board. Have students model each fraction. **Ask:** *How can you rename one fraction so that both have the same denominator?*

2. Have students substitute $\frac{2}{8}$ for $\frac{1}{4}$. **Ask:** *How do you find the sum of two fractions when the denominators are the same?* Have students copy and complete the number sentence.

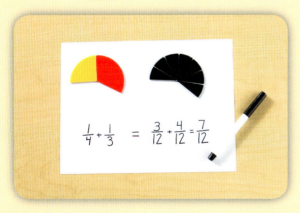

3. Ask: *How would your method change if Jake lived $\frac{1}{3}$ mile from school?* Write $\frac{1}{4} + \frac{1}{3}$ on the board. Have students model $\frac{1}{4} + \frac{1}{3}$ to see that both denominators must be changed before adding. Then have them write an equation for the model.

⚠ Look Out!

Some students may not find a common denominator before adding. These students may add the numerators, then add the denominators. Suggest that they check their solution by using fraction pieces to show the sum, and then place those pieces over the tops of the models for the addends to verify they are an exact match. If they are not an exact match, suggest that students use other same-size fraction pieces to discover a combination that does fit. Students may also need to fit pieces over the original model to find equivalent fractions.

Lesson 12

Number and Operations
Subtract Fractions with Unlike Denominators

Students model the subtraction of fractions with unlike denominators as the first step toward using an algorithm to subtract. Many students will use prior experience with finding equivalent fractions to find common denominators before subtracting. Students may also use number sense or reasoning to find the solution to a problem.

Try It! Perform the Try It! activity on the next page.

Objective
Subtract fractions with unlike denominators.

Skills
- Subtracting rational numbers
- Representing rational numbers
- Finding equivalent fractions

NCTM Expectations
Grades 3–5
Number and Operations
- Recognize and generate equivalent forms of commonly used fractions, decimals, and percents.
- Use visual models, benchmarks, and equivalent forms to add and subtract commonly used fractions and decimals.

Talk About It
Discuss the Try It! activity.

- **Ask:** *When finding the difference between two fractions, why can you subtract the numerators of two fractions with the same denominator, but cannot subtract the numerators of two fractions with unlike denominators?*
- **Ask:** *Which fraction or fractions would you rename to find $\frac{9}{10} - \frac{2}{5}$?*
- Have students compare addition and subtraction of fractions with unlike denominators. **Ask:** *How can you use addition to check your answer to a subtraction problem?*

Solve It
Reread the problem with students. Have students explain in writing why they needed to subtract to find the solution. Then have them describe how they found the quantity of milk needed from the second carton.

More Ideas
For other ways to teach subtracting fractions with unlike denominators—

- Have students use fraction circles or Fraction Tower® Equivalency Cubes to model subtraction problems.
- Have students use the hexagon, blue rhombus, trapezoid, and triangle from a set of pattern blocks to model subtraction. Let the hexagon represent one whole; the blue rhombus, $\frac{1}{3}$; the trapezoid, $\frac{1}{2}$; and the triangle, $\frac{1}{6}$. Have students write as many subtraction sentences as they can, using these fractions. Suggest that students fit pieces over larger pieces to help them find each difference.

Standardized Practice
Have students try the following problem.

Jordan and Mark are painting opposite sides of a fence. Mark has painted $\frac{7}{10}$ of his side. Jordan has painted $\frac{1}{2}$ of his side. How much more has Mark painted than Jordan?

A. $\frac{5}{8}$ B. $\frac{5}{10}$ C. $\frac{1}{5}$ D. $\frac{1}{10}$

Try It! 15 minutes | Groups of 4

Here is a problem about subtracting fractions with unlike denominators.

A cornbread recipe calls for $\frac{3}{4}$ cup of milk. Rachel uses the last $\frac{5}{8}$ cup of milk in one carton. She opens another carton and pours the remaining amount needed. How much milk does Rachel use from the newly opened carton?

Introduce the problem. Then have students do the activity to solve the problem. Distribute fraction squares, paper, and pencils to students. Have students trace 3 whole squares in a row on the paper.

Materials
- fraction squares (2 sets per group)
- paper (11" x 17" 1 sheet per group)
- pencils (1 per group)

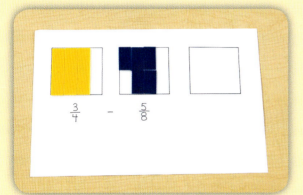

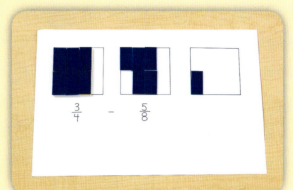

1. Have students model each fraction.
Ask: *What expression can you write to show the situation?* Write $\frac{3}{4} - \frac{5}{8}$ on the board.
Ask: *Using what you know about adding two fractions, what do you think you should do first to subtract these two fractions?*

2. Have students substitute $\frac{6}{8}$ for $\frac{3}{4}$.
Ask: *Now that both fractions have the same denominator, how can you find the difference?* Have students write and model the subtraction equation for the situation.

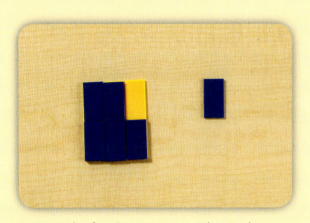

3. Say: *The fraction squares can be used to check your answer.* Have students model each fraction again. This time place the five blue pieces on top of the three yellow pieces.
Ask: *What piece is needed to completely cover the three yellow pieces? What fraction does this piece represent? Is this fraction the same as your answer?*

⚠ Look Out!

Some students may not measure the squares properly when finding equivalent fractions. Suggest that these students place the squares over each other to ensure that the fractional values are exactly the same. Other students may not recognize the importance of finding common denominators, instead looking for a fraction square that will be the same size. Have these students explain why their solution works and how they can use it to find the differences in other situations.

Lesson 13

Number and Operations

Add and Subtract Decimals

Students will deepen their understanding of the base-ten system as they regroup hundredths, tenths, and ones when they add and subtract decimals. Developing number sense helps students grasp the concept of addition and subtraction and leads to an understanding that these two operations are inversely related. This prepares students for multiplying and dividing decimals.

Try It! Perform the Try It! activity on the next page.

Objective
Add and subtract decimals involving tenths and hundredths.

Skills
- Representing decimals
- Adding decimals to hundredths with regrouping
- Subtracting decimals to hundredths with regrouping

NCTM Expectations
Grades 6–8
Number and Operations
- Understand the meaning and effects of arithmetic operations with fractions, decimals, and integers.
- Understand and use the inverse relationships of addition and subtraction, multiplication and division, and squaring and finding square roots to simplify computations and solve problems.

Talk About It
Discuss the Try It! activity.
- **Ask:** *Why do you count the hundredths first?*
- **Ask:** *When adding, why do you need to regroup hundredths? Tenths?*
- **Ask:** *When subtracting hundredths, how do you regroup tenths? How do you regroup ones to subtract tenths?*
- **Ask:** *How can you use inverse operations to check your solutions?*

Solve It
Reread the problem with students. Before students begin to solve the problems, make sure they understand why they are adding or subtracting. Encourage them to check that they have regrouped correctly.

More Ideas
For other ways to teach about adding and subtracting decimals—
- Have students use Fraction Tower® Equivalency Cubes or Snap® Cubes to add and subtract decimals to tenths. When using the Snap Cubes, tell students that each cube represents one tenth and each set of 10 cubes snapped together represents 1 whole. Provide various scenarios using decimals up to 0.9 and then have students add and subtract decimals with regrouping. Have students snap together 10 Equivalency Cubes or Snap Cubes when regrouping 10 tenths as 1 whole to reinforce the relationship between tenths and ones.
- Have students use fraction circles and the decimal fraction circle ring to add and subtract numbers to tenths and hundredths.

Standardized Practice
Have students try the following problem.

Ms. Jenks bought 1.75 pounds of ham and 0.8 pound of turkey from the deli. How many more pounds of ham than turkey did she buy?

A. 2.55 pounds   **C.** 0.95 pound

B. 1.83 pounds   **D.** 1.67 pounds

Try It!   25 minutes | Pairs

Here is a problem about adding and subtracting decimals.

Joaquin hiked 0.56 mile in the morning and 1.45 miles in the afternoon. Joaquin wants to record in his journal how many miles he hiked altogether and how much farther he hiked in the afternoon. What distances should he put in his journal?

Introduce the problem. Then have students do the activity to solve the problem. Distribute base ten blocks to students. Explain to students that a flat equals $\frac{100}{100}$, or 1. Have students determine the value of a rod and of a unit.

Materials
- base ten blocks (2 flats, 20 rods, and 25 units per pair)

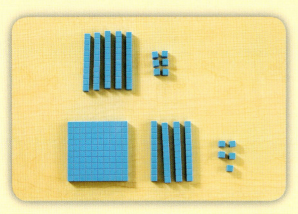

1. Say: *You want to add 0.56 and 1.45 to find the total distance Joaquin hiked.* Have students model 0.56 and 1.45 with the blocks.

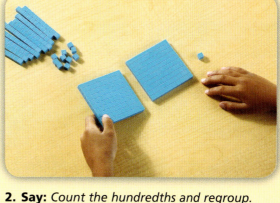

2. Say: *Count the hundredths and regroup. Then count the tenths and regroup.* After students have counted and regrouped, they should show 2 flats and 1 unit. **Ask:** *How many miles did Joaquin hike?*

⚠ Look Out!

Some students might express the solution to the addition problem as 2.1 rather than 2.01. Remind students that 1 unit represents $\frac{1}{100}$ and is expressed 0.01 as a decimal. Model the differences between 2.0 and 2.1 using base ten blocks. Show that 2.0 is the same as 1 whole and 10 tenths and that 2.1 is equivalent to 1 whole and 11 tenths. This will reinforce the importance of placing a zero in the tenths place when regrouping 10 tenths as 1 whole.

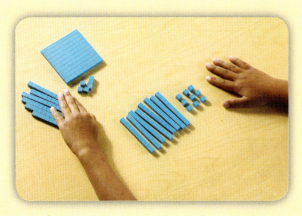

3. Ask: *Now you want to know how much farther Joaquin hiked in the afternoon than the morning. What operation do you use?* Write 1.45 − 0.56 on the board, and have students model the subtraction. Help them regroup when subtracting hundredths and tenths.

Lesson 14

Number and Operations

Multiply with Fractions

As students learn how to multiply with fractions, they continue to use the idea of multiplication as repeated addition. The use of models helps them discover why multiplying with fractions results in a product that is less than one or both factors. This is surprising to students because it is different from multiplying whole numbers where the product is greater than the factors (except when a factor is 0 or 1). Understanding this concept will help them when they multiply decimals.

Try It! Perform the Try It! activity on the next page.

Objective

Multiply with fractions, including a fraction with a whole number and a fraction with a fraction.

Skills

- Representing fractions
- Multiplying with fractions
- Simplifying fractions

NCTM Expectations

Grades 6–8
Number and Operations
- Work flexibly with fractions, decimals, and percents to solve problems.
- Understand the meaning and effects of arithmetic operations with fractions, decimals, and integers.

Talk About It

Discuss the Try It! activity.

- **Ask:** *Why can you use repeated addition to solve this problem?*
- **Ask:** *Why is the product $6 \times \frac{1}{3}$ less than 6?*
- **Ask:** *What size piece is needed to build $\frac{1}{2}$ in $\frac{1}{2} \times \frac{1}{3}$?*

Solve It

Reread the problem with students. Point out that the *of* in $\frac{1}{2}$ of $\frac{1}{3}$ means to multiply, so it can be replaced by the multiplication symbol.

More Ideas

For other ways to teach multiplying with fractions—

- Extend the lesson by using Fraction Tower® Equivalency Cubes to solve this problem: *2 friends share $\frac{3}{4}$ gallon of lemonade. What fraction of the lemonade does each friend drink?* Remind students that $\frac{3}{4}$ is $\frac{1}{4} + \frac{1}{4} + \frac{1}{4}$, so half of $\frac{3}{4}$ is half of $\frac{1}{4}$ taken 3 times. Have students use two $\frac{1}{8}$ pieces to build a tower equal to $\frac{1}{4}$. Have them show that $\frac{1}{8}$ is $\frac{1}{2}$ of $\frac{1}{4}$ and that $\frac{1}{8} + \frac{1}{8} + \frac{1}{8} = \frac{3}{8}$. So, $\frac{1}{2} \times \frac{3}{4} = \frac{3}{8}$.

- Provide different scenarios to give students practice in multiplying fractions and whole numbers, such as $\frac{2}{5} \times 4$ and $\frac{3}{8} \times 3$. Have students use fraction squares to model and solve the problems.

Standardized Practice

Have students try the following problem.

Kelly wants to decrease a recipe by $\frac{2}{3}$. How much less milk should she use if the recipe calls for 6 cups of milk?

A. $\frac{1}{6}$ cup   B. $\frac{1}{3}$ cup   C. 2 cups   D. 4 cups

Try It! 20 minutes | Groups of 4

Here is a problem about multiplying with fractions.

Helen has two hot dogs. She cuts each hot dog into thirds to feed her son, Matt. Matt eats all but one of the pieces. He eats $\frac{1}{2}$ of this remaining piece. How much of a hot dog does Matt not eat?

Introduce the problem. Then have students do the activity to solve the problem. Distribute Fraction Tower® Equivalency Cubes, paper, and pencils to students.

Materials
- Fraction Tower® Equivalency Cubes (2 sets per group)
- paper (1 sheets per group)
- pencils (1 per group)

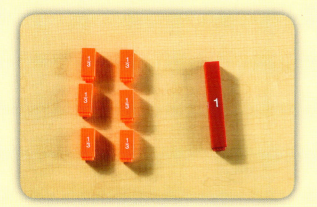

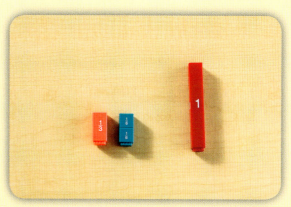

1. Say: *You can use repeated addition to solve the problem. Since there are 6 hot dog pieces, you can add $\frac{1}{3}$ six times.* Have students use six $\frac{1}{3}$ pieces to model the problem. Have students group the pieces and compare them to a whole to determine how many pieces of hot dog matt has.

2. Say: *Matt eats all but one of the pieces. He eats $\frac{1}{2}$ of the remaining piece. Find the fraction that represents the amount of a hot dog Matt did not eat.* Have students place a $\frac{1}{3}$ piece on the table and then build an equivalent tower using two pieces of the same size. **Ask:** *Why does the equivalent tower need to have two pieces?* Be sure students realize that the tower must be equivalent to the $\frac{1}{3}$ piece.

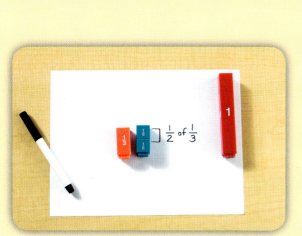

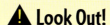

3. Ask: *How can you use the equivalent tower to find $\frac{1}{2}$ of $\frac{1}{3}$?* Have students use a grouping symbol to show that each piece in the tower represents $\frac{1}{2}$ of $\frac{1}{3}$. **Ask:** *What is $\frac{1}{2}$ of $\frac{1}{3}$?*

⚠ Look Out!

Watch for students who compare the $\frac{1}{2}$ piece to the $\frac{1}{3}$ piece when trying to find $\frac{1}{2}$ of $\frac{1}{3}$. Explain that the equivalent tower needs to show $\frac{1}{3}$ divided into halves. To emphasize this connection, have them trace around the $\frac{1}{3}$ piece and then use the divisions in the $\frac{2}{6}$ tower to divide the diagram into halves. Have them shade $\frac{1}{2}$ of the diagram and then compare it to the $\frac{2}{6}$ tower.

Lesson 15

Number and Operations

Meaning of Division

Students explore division as sharing and repeated subtraction, and they should be learning to recognize the relationship between division and multiplication. Both concepts contribute to developing students' number sense, which enables them to understand when and how to use each of the two operations. This lesson focuses on division of whole numbers and prepares students for division of fractions and decimals.

Objective

Explore the meaning of division.

Skills

- Representing whole numbers
- Multiplying whole numbers
- Dividing whole numbers

NCTM Expectations

Grades 3–5
Number and Operations
- Understand various meanings of multiplication and division.
- Understand the effects of multiplying and dividing whole numbers.
- Identify and use relationships between operations, such as division as the inverse of multiplication, to solve problems.
- Develop fluency in adding, subtracting, multiplying, and dividing whole numbers.

Try It! Perform the Try It! activity on the next page.

Talk About It

Discuss the Try It! activity.

- **Ask:** *What is the inverse of repeated subtraction?*
- **Ask:** *Why is distributing among groups to show division called* sharing?
- **Ask:** *How does knowing that multiplication is the inverse of division help you when you are dividing?*

Solve It

Reread the problem with students. Have students compare the models in Steps 1 and 2 and write an equation for each situation.

More Ideas

For other ways to teach about the meaning of division—

- Have students work in groups of 3 or 4 using color tiles to build arrays of different sizes. Students can take turns building arrays and writing division and multiplication sentences to represent them.
- Distribute base ten blocks (10 rods and 20 units per group) to groups of 4 students. Have students model various problems, such as 56 ÷ 4. Students should first model 56 using 5 rods and 6 units, and then divide the rods and units into 4 equal groups. They will need to regroup one rod as 10 units to find the quotient of 14. Present real-world problems to give students practice recognizing when to use division and identifying the meaning of the divisor.

Standardized Practice

Have students try the following problem.

Nikki is displaying 60 photographs in equal rows for a photography contest. If there are 5 rows, how many photographs are in each row?

A. 10   B. 12   C. 14   D. 20

Try It! 20 minutes | Groups of 4

Here is a problem about division of whole numbers.

Two schools are each sending 72 students to a district spelling bee. School A is sending six students from each of their 5th grade classes. How many 5th grade classes does school A have? School B has six 5th grade classes. How many students can be sent from each class?

Introduce the problem. Then have students do the activity to solve the problem. Distribute centimeter cubes, paper, and pencils. Help students recognize this as a division situation.

Materials
- centimeter cubes (72 per group)
- paper (2 sheets per group)
- pencils (1 per group)

1. Say: *You can think about division as repeated subtraction. In this case the divisor tells how many are in a group.* Have students model the 72 students with centimeter cubes and subtract 6 cubes at a time until no cubes remain. **Ask:** *How many times did you subtract 6 cubes?*

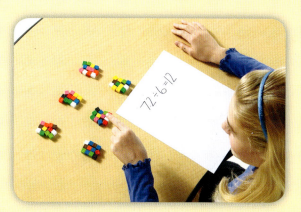

2. Say: *You can also share the cubes among different groups to solve the problem. The divisor tells how many groups.* Have students distribute the 72 cubes among 6 groups until no cubes remain. **Ask:** *How many cubes are in each group?*

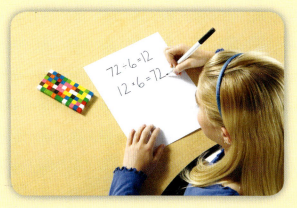

3. Say: *Another way to model the problem is to build an array with 72 cubes. The size of each row is the divisor. The number of rows is the quotient* **Say:** *The array also models multiplication. Write a division and multiplication sentence for the array.*

⚠ Look Out!

Some students may have difficulty forming groups or building arrays. Students can draw 6 circles on the sheet of paper for Step 2, and then place one cube at a time in each circle until all the cubes have been used. For Step 3, encourage students to line up the rows and columns as they build each row so they can easily see the number of rows and columns.

Lesson 16

Number and Operations

Divide with Fractions

When students develop their understanding of division with whole numbers, they make the connection that the divisor is either the number of groups to form or the number in each group and that the dividend is the number that is divided into groups. Students can apply the same model when they divide with fractions. They also understand that dividing a fraction by a number is the same as multiplying the fraction by the reciprocal of the number.

Try It! *Perform the Try It! activity on the next page.*

Objective

Divide with fractions.

Skills

- Representing rational numbers
- Dividing fractions
- Multiplying fractions

NCTM Expectations

Grades 6–8
Number and Operations
- Work flexibly with fractions, decimals, and percents to solve problems.
- Understand the meaning and effects of arithmetic operations with fractions, decimals, and integers.

Talk About It

Discuss the Try It! activity.

- **Ask:** *What is the divisor in the problem? What is the dividend?*
- **Ask:** *Why can you think of the problem as half of $\frac{3}{4}$?*
- **Ask:** *Is $\frac{3}{4} \div 2$ the same as $2 \div \frac{3}{4}$? How can you tell?*

Solve It

Reread the problem with students. Some students may readily see that $\frac{6}{8}$ is equivalent to $\frac{3}{4}$ and $\frac{6}{8}$ is two groups of $\frac{3}{8}$, while others may need to try out various fraction pieces. If time permits, explain that $\frac{1}{2}$ is the reciprocal of $\frac{2}{1}$ and show that a number multiplied by its reciprocal is equal to 1.

More Ideas

For other ways to teach dividing with fractions—

- Use fraction circles to model division with fractions such as $\frac{1}{3} \div 4$. Have students start with an orange $\frac{1}{3}$ piece and then find four same-sized pieces that fit exactly on the orange piece. Students find that four black pieces fit exactly on the orange piece, so, $\frac{1}{3} \div 4$ is $\frac{1}{12}$.
- Extend the lesson by using Fraction Tower® Equivalency Cubes to divide a fraction by a fraction, such as $\frac{2}{3} \div \frac{1}{6}$. Tell students that they want to determine how many sixths are in $\frac{2}{3}$. Have students build a tower of two one-third pieces and an equal tower of one-sixth pieces to show there are four one-sixth pieces in $\frac{2}{3}$, which means that $\frac{2}{3} \div \frac{1}{6}$ is 4.

Standardized Practice

Have students try the following problem.

If Liam cuts $\frac{1}{2}$ yard of rope into 3 equal pieces, what is the length of each piece?

A. $\frac{1}{12}$ yard B. $\frac{1}{6}$ yard C. $\frac{1}{3}$ yard D. $3\frac{1}{2}$ yard

50

Try It! 40 minutes | Groups of 4

Here is a problem about dividing with fractions.

Megan has $\frac{3}{4}$ yard of green felt to use on two bulletin boards. She wants to use the same amount of felt on each board. How much felt will she use on each bulletin board?

Introduce the problem. Then have students do the activity to solve the problem. Distribute fraction squares, worksheets, and pencils to students. Review division as sharing.

Materials
- fraction squares (1 set per group)
- Eighths Fraction Squares worksheet (BLM 7)
- pencils (1 per group)

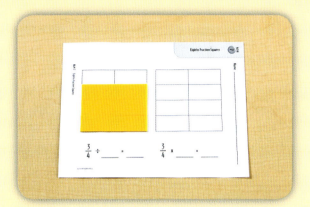

1. Say: *One way to think of the problem is that $\frac{3}{4}$ is divided into two equal parts. Use your fraction squares to model $\frac{3}{4}$.* Have students place three one-fourth yellow pieces on the worksheet.

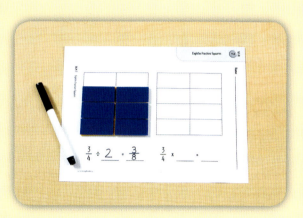

2. Say: *Find two equal fraction pieces that will cover a one-fourth piece.* Have students use blue pieces to cover the yellow pieces. **Ask:** *What size are the new pieces?* **Say:** *Separate the eighths into two equal groups.* **Ask:** *How many eighths are in each group?* Have students complete the division equation beneath the model.

⚠ Look Out!

Some students may try a mixture of fraction pieces to model the problem. Tell them they must use the same color of fraction pieces since they want to form equal groups. If students do not realize that the blue pieces are eighths, have them assemble a fraction square out of blue pieces and then count the pieces.

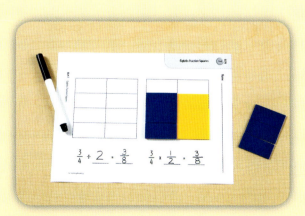

3. Say: *Another way to think of this problem is that half of $\frac{3}{4}$ yard is used for each bulletin board.* Have students place a blue piece on the left half of each yellow to show that $\frac{3}{8}$ is half of $\frac{3}{4}$. **Ask:** *What operation do you use to find half of $\frac{3}{4}$?* Have students complete the multiplication equation beneath the model.

Lesson 17

Number and Operations

Multiply and Divide Decimals

As students multiply and divide decimals, they use and develop concepts such as multiplication as repeated addition and division as sharing. These are concepts that students learned when they worked with whole numbers and fractions. After completing this activity, students should know when to add, subtract, multiply, or divide to solve a real-word decimal problem.

Try It! Perform the Try It! activity on the next page.

Objective
Multiply and divide decimals to hundredths.

Skills
- Representing decimals
- Multiplying decimals
- Dividing decimals

NCTM Expectations
Grades 6–8
Number and Operations
- Work flexibly with fractions, decimals, and percents to solve problems.
- Understand the meaning and effects of arithmetic operations with fractions, decimals, and integers.

Talk About It
Discuss the Try It! activity.

- **Ask:** *How does repeated addition help you model the problem?*
- **Ask:** *In Step 2, why do you regroup the units first and then the rods?*
- **Ask:** *In Step 4, why do you regroup in the opposite order that you regrouped in Step 2, from flats to rods instead of from units to rods?*

Solve It
Reread the problem with students. Have them count the total number of rods and units after they regroup. Remind them to regroup 10 units as one rod, 10 rods as one flat, and one flat as 10 rods as necessary.

More Ideas
For other ways to teach multiplying and dividing decimals—

- Have students use base ten blocks and a Hundredths Grid (BLM 9) to multiply a decimal by a whole number, such as 0.4×2. Have them model 0.4 using 4 rows of a Hundredths Grid (BLM 9), remove the rods, and shade the rows in one color. Then have them repeat the model below the first model. Have students count the tenths.

- Have students use base ten blocks to divide a decimal by a whole number, such as $1.5 \div 3$. Have them model 1.5 and then regroup the one flat as 10 rods. Now have them use repeated subtraction to group 3 rods at a time until no rods remain.

Standardized Practice
Have students try the following problem.

Katrina can swim 5 meters in 4.35 seconds. How long does it take her to swim one meter?

A. 0.81 second **C.** 0.95 second
B. 0.87 second **D.** 0.91 second

Try It! 30 minutes | Groups of 4

Here is a problem about multiplying and dividing decimals.

Diego bought 3 DVDs from an online store. If one DVD weighs 0.48 pound, what is the shipping weight of 3 DVDs? Diego also bought 3 books online. The total shipping weight for the books is 2.37 pounds. What is the shipping weight of one book?

Introduce the problem. Then have students do the activity to solve the problem. Distribute base ten blocks, paper, and pencils. Tell students that a flat represents 1.0. Have students determine what a rod and a unit represent.

Materials
- base ten blocks (2 Flats, 20 rods, and 30 units per group)
- paper (11" x 17", 2 sheets per group)
- pencils (1 per group)

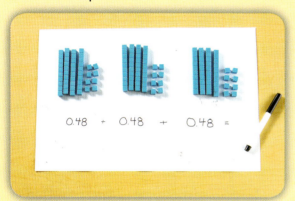

1. Say: *Multiplication is repeated addition, so you can model the problem as 0.48 + 0.48 + 0.48.* Have students model the problem as 3 sets of 4 rods and 8 units and write 0.48 + 0.48 + 0.48 = on the first sheet of paper.

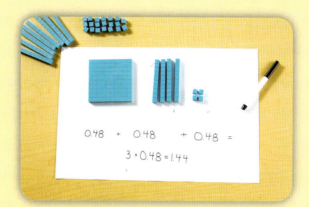

2. Say: *Combine the rods and units.* Have students group the rods and group the units. **Say:** *Regroup the units as rods and the rods as flats.* **Ask:** *What is the total shipping weight of the DVDs?* Have students write $3 \times 0.48 = 1.44$ below the first equation.

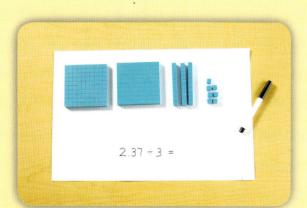

3. Say: *The total shipping weight of three books is 2.37 pounds.* **Ask:** *What is the shipping weight of one book? What do you need to do to solve the problem?* Have students use the base ten blocks to model 2.37 and write 2.37 ÷ 3 = on the second sheet of paper.

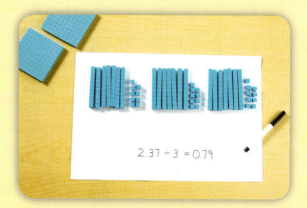

4. Say: *To form 3 equal groups, you must first regroup the flats as rods and then share them equally among the groups.* Have students form the 3 equal groups. **Say:** *Now regroup the rods as units and share them equally among the groups.* **Ask:** *What is the weight of one book?* Finally, have students complete the equation 2.37 ÷ 3 = 0.79.

Number and Operations

Lesson 18

Number and Operations

Ratios

Students use ratios to show various relationships between quantities, including whole to part, part to whole, and part to part. With an understanding of ratio, students can engage in proportional reasoning, which is a major component of a student's foundation in math.

Objective
Use ratios to represent relationships.

Skills
- Representing ratios
- Identifying ratios
- Reasoning

NCTM Expectation
Grades 6–8
Number and Operations
- Understand and use ratios and proportions to represent quantitative relationships.

Try It! Perform the Try It! activity on the next page.

Talk About It
Discuss the Try It! activity.

- **Say:** *Each type of animal represents a part. All the parts together, or the total number of animals, is the whole.* **Ask:** *How many animals are at the habitat?*
- **Say:** *A ratio shows the order of the comparison, so when you compare animals to snakes, the first term is the number of animals and the second term is the number of snakes.*

Solve It
Reread the problem with students. As they represent and identify the ratios, have them write the ratios in words and in numbers. Help them identify whether the ratio is whole to part, part to whole, or part to part.

More Ideas
For other ways to teach ratios—

- Use two-color counters to solve problems such as *The ratio of frogs to ducks in a pond is 5 to 3. Write three ratios to represent this situation.* Have students represent the frogs with the yellow side of the counters and the ducks with the red side of the counters. **Say:** *The total number of ducks and frogs is 8, so the ratio of frogs to the total number of ducks and frogs is 5 to 8.* Guide students to write other ratios.

- Extend the lesson using Cuisenaire® Rods to generate equal ratios. Provide a scenario such as *A trail mix recipe calls for 1 cup of raisins and 3 cups of peanuts. Find the amounts of raisins and peanuts in three different-size batches of this recipe.* Have students build trains on the Centimeter Grid (BLM 8) for each ratio, using white rods for raisins and light green rods for peanuts. Guide students to use multiplication to find other ratios.

Standardized Practice
Have students try the following problem.

There are 3 parrots, 7 parakeets, and 2 finches at a pet store. What is the ratio of parakeets to birds?

A. 12 to 7 **B.** 7:5 **C.** $\frac{7}{12}$ **D.** 5 to 7

Try It! 20 minutes | Groups of 4

Here is a problem about ratios.

An animal habitat includes 3 snakes, 2 alligators, and 5 lizards. What are three ratios you can use to describe the relationships between these animals?

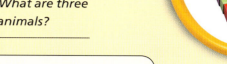

Introduce the problem. Then have students do the activity to solve the problem. Distribute the Cuisenaire® Rods, grid paper, paper, and pencils to students. Write *a to b*, *a:b*, and $\frac{a}{b}$ on the board. Explain that ratios are expressed in these three ways. Ratios compare wholes to parts, parts to wholes, and parts to parts.

Materials
- Cuisenaire® Rods (1 set per group)
- Centimeter Grid Paper (BLM 8, 1 per group)
- paper (1 sheet per group)
- pencils (1 per group)

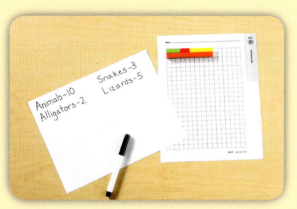

1. Say: *Find rods to represent the number of each type of animal. Place the rods in a train on the first row of the grid.* **Say:** *Now find a rod to represent the total number of animals. Place this rod on the second row.* Have them write *Animals -10, Snakes - 3, Alligators - 2, Lizards - 5* on the paper.

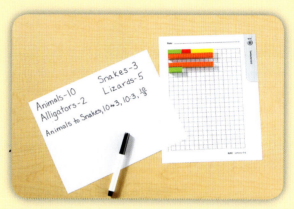

2. Say: *Compare the total number of animals to the number of snakes.* Have students place the orange rod on one row and the green rod on the next row. **Say:** *This ratio compares the whole to a part.* **Ask:** *What is the ratio of animals to snakes?* Have students write the ratio three ways.

3. Say: *Compare lizards to animals. Express the ratio in three ways. This ratio compares a part to the whole. Compare lizards to alligators, alligators to snakes, and snakes to lizards. Express each ratio three ways.* **Ask:** *What do these ratios compare?*

⚠ Look Out!

Some students may write 7 to 3 or 5 to 5 when writing the ratios for animals to snakes or lizards to animals. Stress that snakes and lizards are parts of the total number of animals, so they must include them in the whole when comparing a whole to a part or a part to a whole. Have students write ratios for other whole-to-part or part-to-whole comparisons to make sure they understand the concept.

Lesson 19

Number and Operations

Proportions

Students use their knowledge of ratios to represent and solve proportions. They use models to find equivalent ratios and use ratios to solve a proportion. They learn that a proportion is a statement that two ratios are equal. These activities develop proportional thinking, which students use to solve problems involving rates, unit conversions, and functions.

Try It! Perform the Try It! activity on the next page.

Objective
Use proportions to represent relationships.

Skills
- Representing proportions
- Solving proportions
- Reasoning

NCTM Expectations

Grades 6–8
Number and Operations
- Understand and use ratios and proportions to represent quantitative relationships.
- Develop, analyze, and explain methods for solving problems involving proportions, such as scaling and finding equivalent ratios.

Talk About It

Discuss the Try It! activity.

- **Say:** *You can write proportions two ways. For example, you can express 3 hearts to 2 diamonds down or 3 hearts to 2 diamonds across.* Circle *3 to 2* down and *3 to 2* across in the proportions written on the board. Explain that whichever form they choose, the two ratios must be constructed in the same order.

- **Ask:** *Why do you keep adding pairs until you reach 15 hearts? Can you use any equal ratio to solve the proportion?*

- **Ask:** *What pattern do you recognize as you find equal ratios? What do you think proportion means?*

Solve It

Reread the problem with students. Have students write each proportion represented by the models. Emphasize that only one solution works for the problem, but that any equal ratios define a proportion.

More Ideas

For other ways to teach proportions—

- Have students practice using equal ratios to form proportions. Give them a ratio such as 3 dogs to 5 cats. Have them work in pairs using centimeter cubes to generate the equal ratios. Have them form proportions and write them in two different ways.

- Have students generate equal ratios in tables and use the ratio tables to form proportions. Students work in pairs using two-color counters to form the equal ratios.

Standardized Practice

Have students try the following problem.

The ratio of benches to trees in a park is 2:9. If there are 18 trees, how many benches are in the park?

A. 4 benches   **C.** 14 benches

B. 7 benches   **D.** 9 benches

Try It! 30 minutes | Groups of 4

Here is a problem about proportions.

A company makes charms for bracelets. For every 3 hearts, it makes 2 diamonds. If the company makes 15 hearts, how many diamonds does it make?

Introduce the problem. Then have students do the activity to solve the problem. Distribute Cuisenaire® Rods, paper, and pencils. Explain that a proportion is formed by two equivalent ratios. Write on the board:

$$\frac{3 \text{ hearts}}{2 \text{ diamonds}} = \frac{15 \text{ hearts}}{? \text{ diamonds}}$$

Materials
- Cuisenaire® Rods (1 set per group)
- paper (2 sheets per group)
- pencils (1 per group)

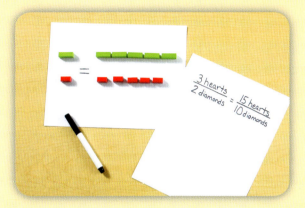

1. Say: *Equal ratios can help you solve a proportion.* Have students write an equal sign on a sheet of paper. **Say:** *Place a pair of rods to the left of the equal sign to represent 3 hearts: 2 diamonds. Build an equal ratio to the right by adding 3 hearts:2 diamonds until you reach 15 hearts. Count the diamonds. Write the proportion.*

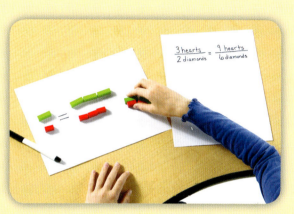

2. Say: *You can form a proportion out of any two equal ratios.* Have students remove two red and two green rods from their model on the right side of the equal sign. **Ask:** *What proportion does the model represent now? How do you know it is a proportion?* Have students build another proportion.

⚠ Look Out!

Some students might have trouble writing the proportion correctly. Emphasize that the comparison is down or across. Set up a framework for students to fill in:

$$\frac{\square \text{ hearts}}{\square \text{ diamonds}} = \frac{\square \text{ hearts}}{\square \text{ diamonds}}$$

or

$$\frac{\square \text{ hearts}}{\square \text{ hearts}} = \frac{\square \text{ diamonds}}{\square \text{ diamonds}}$$

Have them fill in 3:2 down and 3:2 across to emphasize the pattern.

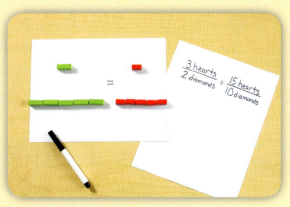

3. Have students model and solve the proportion

$$\frac{3 \text{ hearts}}{15 \text{ hearts}} = \frac{2 \text{ diamonds}}{? \text{ diamonds}}$$

Have them compare their answer with their answer from Step 1.

Geometry

Geometry at this level builds on the basic concepts of shape that were addressed at previous levels. Emphasis is placed on the properties of geometric figures and their relationships to one another. Students develop a stronger awareness of the role that geometry plays in the world around them.

Students classify shapes as they have previously, but they take this activity to new levels. They apply a deeper knowledge of angles and relationships between sides, and they more readily apply the concepts of congruence and symmetry. They cover the description and classification of plane shapes thoroughly and give substantial attention to solids as well.

> **The NCTM Standards for Geometry suggest that students should:**
> - Analyze characteristics and properties of two- and three-dimensional geometric shapes and develop mathematical arguments about geometric relationships
> - Specify locations and describe spatial relationships using coordinate geometry and other representational systems
> - Apply transformations and use symmetry to analyze mathematical situations
> - Use visualization, spatial reasoning, and geometric modeling to solve problems

In Grades 5 and 6, students learn to use coordinate geometry routinely and for various purposes. Coordinate systems are powerful tools for defining shapes and for describing the properties of shapes and the relationships between shapes. Using coordinate systems, students can attach quantitative meaning to transformations and symmetry and thereby strengthen their understanding of these concepts. Coordinate geometry is also useful for demonstrating perpendicular and parallel lines, congruence, and similarity.

In these grades, students start to use mental constructions, drawings, and algebraic processes to express geometric concepts. To do this, students must form accurate spatial visualizations. This can be a difficult task, but students will find aid in the frequent handling of concrete models. The following activities are built around manipulatives that students can use to develop skills and explore concepts in **Geometry**.

Geometry
Contents

Lesson 1 Measure and Classify Angles 60
  Objective: Recognize types of angles.
  Manipulative: geoboard

Lesson 2 Identify and Classify Triangles 62
  Objective: Identify and classify triangles.
  Manipulative: AngLegs™

Lesson 3 Identify and Classify Quadrilaterals .. 64
  Objective: Identify and classify quadrilaterals.
  Manipulative: AngLegs™

Lesson 4 Regular Polygons 66
  Objective: Identify and classify regular polygons.
  Manipulative: AngLegs™

Lesson 5 Line Symmetry 68
  Objective: Identify and draw lines of symmetry in polygons.
  Manipulative: pattern blocks and GeoReflector™ mirror

Lesson 6 Parallel and Perpendicular Lines 70
  Objective: Identify parallel and perpendicular lines.
  Manipulative: geoboard

Lesson 7 Shapes in the Coordinate Plane 72
  Objective: Draw shapes on a coordinate grid and describe their properties.
  Manipulative: AngLegs™

Lesson 8 Slides and Flips 74
  Objective: Identify and describe slides and flips.
  Manipulative: AngLegs™

Lesson 9 Rotational Symmetry 76
  Objective: Identify whether figures have rotational symmetry.
  Manipulative: GeoReflector™ mirror and pattern blocks

Lesson 10 Multiple Transformations 78
  Objective: Perform multiple transformations on geometric figures.
  Manipulative: AngLegs™

Lesson 11 Tessellations 80
  Objective: Investigate the characteristics of tessellations.
  Manipulative: pattern blocks

Lesson 12 Congruent Figures and Transformations 82
  Objective: Identify congruent figures.
  Manipulative: AngLegs™

Lesson 13 Corresponding Parts of Congruent Figures 84
  Objective: Identify corresponding parts of congruent figures.
  Manipulatives: AngLegs™

Lesson 14 Similar Triangles 86
  Objective: : Build similar triangles.
  Manipulative: AngLegs™

Lesson 15 Nets 88
  Objective: Explore nets.
  Manipulative: Relational Geosolids®

Lesson 16 Three-Dimensional Shapes 90
  Objective: Classify three-dimensional shapes.
  Manipulative: Relational Geosolids®

Lesson 1

Geometry
Measure and Classify Angles

Students need a basic understanding of angles to learn the properties of two-dimensional shapes. In this lesson, students use models to represent, measure, and classify angles.

Try It! *Perform the Try It! activity on the next page.*

Objective
Recognize types of angles.

Skills
- Drawing angles
- Comparing angles
- Recognizing angles in everyday situations

NCTM Expectations

Grades 3–5
Geometry
- Identify, compare, and analyze attributes of two- and three-dimensional shapes and develop vocabulary to describe the attributes.
- Build and draw geometric objects.
- Recognize geometric ideas and relationships and apply them to other disciplines and to problems that arise in the classroom or in everyday life.

Talk About It
Discuss the Try It! activity.
- Have students define an angle in terms of its sides and vertex.
- **Ask:** *There are two scales on a protractor, how do you know which scale to use?*
- **Ask:** *How can you use a right angle to help you classify angles?*

Solve It
Reread the problem with students. Have them classify each angle. Then have them explain how to draw and measure angles.

More Ideas
For other ways to teach about measuring and classifying angles—
- Have students make the initials of their first and last names on geoboards. Then have them measure and record the angles in their initials. Encourage students to compare their results and note the differences in the types of angles associated with a letter.
- Have students work in pairs. Using various colors of AngLegs™, have one student construct an angle and the other measure and then classify it. Have students use a variety of orientations for the angles.

Standardized Practice
Have students try the following problem.

Which best describes the angle at the intersection of Brown Street and Ivy Road?

A. acute   C. right

B. obtuse   D. straight

Ivy Rd.

Brown St.

Try It! 30 minutes | Groups of 4

Here is a problem about measuring and classifying angles.

The figure shows the route Mr. Chen takes to work— Oak to Pine, Pine to Willow, and Willow to Olive. What type of angle is created by the streets at Mr. Chen's first turn? His second? His third?

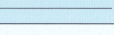

Introduce the problem. Then have students do the activity to solve the problem. Distribute geoboards, rubber bands, AngLegs™, protractors, paper, and pencils. Draw the figure on the board. **Say:** *This is Mr. Chen's route. Oak St. starts at the lower left corner.*

Materials
- geoboards (1 per group)
- rubber bands (4 per group)
- AngLegs™ (1 set per group)
- protractors (2 per group, included in AngLegs™)
- paper (2 sheets per group)
- pencils (1 per group)

1. Say: *Angles can be classified as acute, obtuse, right, or straight.* Demonstrate the types of angles using AngLegs. Point out the sides and vertex of an angle. **Say:** *Model each type of angle using AngLegs, trace the angle on a sheet of paper, and label it.*

2. Say: *Measure the angles you drew.* Show how to align the protractor with the vertex and sides of the angle. **Say:** *If an angle opens to the right, use the lower scale. If it opens to the left, use the upper scale.* Have students measure their angles.

⚠ Look Out!

Students may use the wrong scale when measuring angles. Encourage them to first classify the angle, and then check whether the measure of the angle fits its classification. If it does not, have them reread the scale. If students find it difficult to classify the angles on the geoboard, have them lay AngLegs on top of the angle, and then trace it onto a sheet of paper and measure it with the protractor.

3. Have students make Mr Chen's route on a geoboard and have them classify the angles in the route. They can use a right angle made from AngLegs as a reference.

LESSON 2

Geometry

Identify and Classify Triangles

Classifying triangles helps students develop reasoning skills that they will use when they study similar and congruent triangles. By building triangles, students visualize possible and impossible combinations of angles and side lengths. They also can learn to make generalizations about the properties of triangles.

Try It! Perform the Try It! activity on the next page.

Talk About It

Discuss the Try It! activity.

- **Ask:** *Is it possible to build a triangle with two right angles? Two obtuse angles? A right angle and an obtuse angle? What do you notice when you build a triangle that has a right or obtuse angle?*
- **Ask:** *Looking at the equilateral, isosceles, and scalene triangles, can you suggest any generalizations about these triangles?*

Solve It

Reread the problem with students. Have students describe the triangles they built and discuss the different ways students could sort the triangles.

More Ideas

For other ways to teach about identifying and classifying triangles—

- Have students work in pairs. One student makes a triangle on a geoboard and the other classifies the triangle by its angles and its sides.
- Draw a shape on the board, such as a rectangle bisected by two diagonals. Ask students to duplicate the shape on the geoboard and find as many triangles as they can. Have students classify the triangles.

Standardized Practice

Have students try the following problem.

Which best describes the triangle?

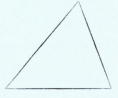

A. acute, equilateral

B. obtuse, isosceles

C. acute, scalene

D. obtuse, scalene

Objective

Identify and classify triangles.

Skills

- Classifying angles
- Classifying triangles
- Reasoning

NCTM Expectations

Grades 3–5
Geometry
- Identify, compare, and analyze attributes of two- and three-dimensional shapes and develop vocabulary to describe the attributes.
- Classify two- and three-dimensional shapes according to their properties and develop definitions of classes of shapes such as triangles and pyramids.
- Build and draw geometric objects.

Try It! 25 minutes | Groups of 4

Here is a problem about identifying and classifying triangles.

Andrew wants to make a quilt using triangles. How many types of triangles could Andrew use?

Introduce the problem. Then have students do the activity to solve the problem. Distribute AngLegs™, protractors, recording charts, and pencils. Explain that all triangles have three sides and three angles. **Say:** *Triangles can be classified by their angles, their sides, or both.* Have students start two charts with the headings: *Name of Triangle, Angles,* and *Sketch* on one chart and *Name of Triangle, Sides,* and *Sketch* on the other to record their results.

Materials
- AngLegs™ (1 set per group)
- 4-Column Recording Chart (BLM 17; 3 per group)
- pencils (1 per group)

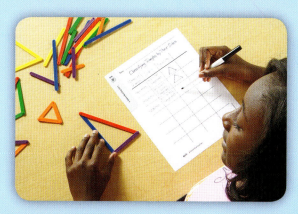

1. Say: *Triangles can be classified by their angles as acute, obtuse, or right.* Review the types of angles. Have students build a right triangle they can use to judge the angles in other triangles. **Say:** *Build examples of each type of triangle and record their properties.*

2. Say: *Triangles can be classified by their sides as equilateral, isosceles, or scalene.* Describe the triangles according to the number of congruent sides. Have students build and record each type of triangle. **Ask:** *Is an equilateral triangle also isosceles?*

3. Say: *Build all the different types of triangles classified by their angles and sides, for example acute scalene.* Have students start a third chart to record their findings. **Ask:** *How many types of triangles can you build?*

⚠ Look Out!

Some students may have difficulty building and sorting triangles classified by both angles and sides. Suggest that they start with the first type of angle on their recording sheet and then build as many triangles as possible using that type of angle and the three different types of sides. Then have them repeat this for each of the other types of angles.

Lesson 3

Geometry

Identify and Classify Quadrilaterals

In this lesson, students investigate the properties of quadrilaterals by making models. They learn that some quadrilaterals can be classified in more than one way. Knowing the properties of quadrilaterals prepares students to find area and volume.

Objective

Identify and classify quadrilaterals.

Skills

- Classifying angles
- Recognizing parallel lines
- Reasoning

NCTM Expectations

Grades 3–5
Geometry
- Identify, compare, and analyze attributes of two- and three-dimensional shapes and develop vocabulary to describe the attributes.
- Classify two- and three-dimensional shapes according to their properties and develop definitions of classes and shapes such as triangles and pyramids.
- Build and draw geometric objects.

Try It! Perform the Try It! activity on the next page.

Talk About It

Discuss the Try It! activity.

- **Ask:** *Why can neither student make a trapezoid?*
- **Ask:** *What type of AngLeg™ would you need to make a trapezoid?* Ask: *Which quadrilaterals have all right angles?*

Solve It

Reread the problem with students. Have them sketch each of the quadrilaterals and justify why they can or cannot make the quadrilaterals given each set of AngLegs. Have students compare their sketches.

More Ideas

For other ways to teach about identifying and classifying quadrilaterals—

- Have students use pattern blocks to identify and classify as many quadrilaterals as possible. Challenge them to combine shapes to build any type of quadrilateral that is missing from the pattern blocks set.
- Have students work in pairs. The first student describes at least three characteristics of a quadrilateral and the other student makes the quadrilateral on a geoboard. The two students classify the quadrilateral in as many ways and as specifically as possible.

Standardized Practice

Have students try the following problem.

Which best describes the quadrilateral?

A. parallelogram

B. rectangle

C. rhombus

D. square

Try It! 25 minutes | Pairs

Here is a problem about identifying and classifying quadrilaterals.

Owen and Lili are using AngLegs™ to model shapes for an art project. Owen has 2 purple and 2 orange AngLegs. Lili has 4 green AngLegs. How many quadrilaterals can each student make?

Introduce the problem. Then have students do the activity to solve the problem. Distribute AngLegs™, charts, and pencils to students. **Say:** *Figures with four sides and four angles are quadrilaterals.*

Materials
- AngLegs™ (2 purple, 2 orange, and 4 green per pair)
- Quadrilaterals Chart (BLM 15; 2 per pair)
- paper (2 sheets per pair)
- pencils (2 per pair)

1. Tell students to refer to the chart and note the characteristics of a trapezoid. Then have them see whether they can build a trapezoid. Tell students to mark the appropriate column — yes or no — at the bottom of the sheet.

2. Instruct students to repeat this process for each of the shapes listed in the chart. Have students tally the yes and no responses.

3. Have students note the similarities that exist among some of the shapes. Discuss, for example, that a square is a special rectangle, and that a rhombus is a special kite.

⚠ Look Out!

Some students may think that since any rhombus, rectangle, or square is a parallelogram that any parallelogram must also be a rhombus, a rectangle, and a square. Guide students to see that by definition a square, rectangle, and rhombus have special characteristics.

Lesson 4

Geometry
Regular Polygons

Once students can identify and classify triangles and quadrilaterals by the measures of their sides and by their angles, they can explore the properties of regular polygons. Knowing the properties of regular polygons will help students understand the concepts of congruency and similarity.

Objective
Identify and classify regular polygons.

Skills
- Recognizing attributes of polygons
- Reasoning

NCTM Expectations
Grades 6–8
Geometry
- Precisely describe, classify, and understand relationships among types of two- and three-dimensional objects using their defining properties.
- Draw geometric objects with specified properties, such as side lengths or angle measures.
- Recognize and apply geometric ideas and relationships in areas outside the mathematics classroom, such as art, science, and everyday life.

Try It! Perform the Try It! activity on the next page.

Talk About It
Discuss the Try It! activity.

- **Ask:** *Is a right triangle a regular polygon? A rhombus? A rectangle? A trapezoid? Why is it not possible to classify these polygons as regular polygons?*
- **Ask:** *When you divide a regular polygon into triangles, how do you know that you have formed as many triangles as possible?* **Say:** *Remember the diagonals all start from one vertex.*
- **Ask:** *Can you see a pattern in the number of sides that a polygon has and the number of triangles that can be formed from one vertex? How can you state this as a rule?*

Solve It
Reread the problem with students. Have them sketch or trace the model of the hexagon and explain why the floor of the gazebo is a hexagon.

More Ideas
For other ways to teach about regular polygons—

- Extend the lesson by having students work in pairs using AngLegs™ to build regular polygons with 7, 8, 9, and 10 sides. Have them draw each polygon and predict the measure of each internal angle and the sum of the angles using what they learned in the lesson. Tell students to test their predictions by dividing the polygons into triangles using a pencil and a straightedge.
- Give pairs of students pattern blocks. Have them sort the blocks into two groups—regular polygons and irregular polygons. Have students verify their classification for each type of polygon by using rulers to measure the sides of the polygon and protractors to measure the angles.

Standardized Practice
Have students try the following problem.

Each angle in a regular polygon is 108°. Which polygon is it?

A. square **B.** pentagon **C.** hexagon **D.** octagon

Try It! 30 minutes | Groups of 4

Here is a problem about classifying regular polygons.

The floor of a gazebo has the shape of a regular polygon. Each angle of the polygon measures 120°. What polygon describes the shape of the floor?

Introduce the problem. Then have students do the activity to solve the problem. Distribute AngLegs™, recording charts, paper, and pencils to students.
Say: *A regular polygon is a closed figure with all sides the same length and all angles the same measure.* Have students write the headings *Name, Number of Sides, Number of Triangles, Sum of the Angle Measures,* and *Measure of Each Angle* on their charts.

Materials
- AngLegs™ (1 set per group)
- 6-Column Recording Chart (BLM 18; 1 per group)
- paper (1 sheet per group)
- pencils (1 per group)

1. Say: *Build a 3-sided polygon with blue AngLegs and a 4-sided polygon with yellow AngLegs.* **Ask:** *What are the names of these polygons? Are they regular?* Have students build 5-sided and 6-sided polygons using green and purple AngLegs, respectively, and name each polygon on the chart.

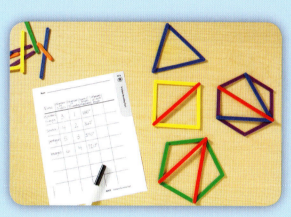

2. Say: *You can divide a polygon with more than three sides into triangles. Choose a vertex. Connect a red or blue AngLeg from that vertex to each of the other vertices.* Demonstrate to students how the sum of the internal angles is equal to 180° multiplied by the number of triangles formed. Have students find the sums of the internal angles of the polygons.

⚠ Look Out!

If students make models in step 1 that look like irregular polygons, then draw a regular pentagon and hexagon on the board and have students correct their models. The diagonals make the polygons rigid and enable students to count the triangles. Remind students that the angle measures in a regular polygon are equal.

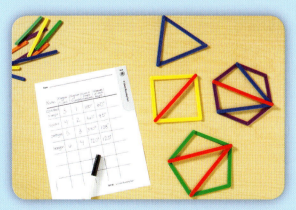

3. Say: *To find the measure of each angle in a regular polygon, divide the sum of its angle measures by the number of sides it has.* Have students complete the recording chart and determine the shape of the gazebo floor.

Lesson 5

Geometry

Line Symmetry

Identifying lines of symmetry in polygons helps students recognize symmetry in the world around them. Line symmetry is the basis of reflection, one of the transformations used in geometric thinking. The concept of line symmetry complements the concepts of congruence and similarity.

Try It! Perform the Try It! activity on the next page.

Objective
Identify and draw lines of symmetry in polygons.

Skills
- Identifying polygons
- Identifying symmetry
- Reasoning

NCTM Expectations

Grades 3–5 Geometry
- Identify and describe line and rotational symmetry in two- and three-dimensional shapes and designs.
- Build and draw geometric objects.
- Create and describe mental images of objects, patterns, and paths.
- Recognize geometric ideas and relationships and apply them to other disciplines and to problems that arise in the classroom or in everyday life.

Talk About It
Discuss the Try It! activity.
- **Ask:** *Are all lines that divide a shape in half called lines of symmetry?*
- **Ask:** *Why is a line of symmetry sometimes called a mirror line or a line of reflection?*
- **Ask:** *How many lines of symmetry does the hexagon have?*
- **Say:** Three of the lines of symmetry divide the hexagon into trapezoids.
  Ask: *What shapes is the hexagon divided into by the other lines of symmetry?* Students should see that the shapes are irregular pentagons.

Solve It
Reread the problem with students. Have them write a paragraph to explain what they know about line symmetry and how they used the mirror to find lines of symmetry.

More Ideas
For other ways to teach line symmetry—
- Have students use pattern blocks and a GeoReflector™ mirror to construct more shapes using the same method they used to construct the hexagon in Step 2 of the Try It! activity. Have them draw the shapes that they create.
- Have students create figures with line symmetry using AngLegs™. Students can make two halves separately and snap the halves together to make a symmetrical shape.

Standardized Practice
Have students try the following problem.

Which figure has three lines of symmetry?

A.   B.   C.   D.

Try It! 30 minutes | Groups of 4

Here is a problem about line symmetry.

Kim's company makes pattern blocks. They make trapezoid blocks by cutting hexagon blocks in half. How many ways can a hexagon block be cut in half to make a pair of trapezoid blocks? Consider other lines that divide the hexagon into halves. In all, how many different lines produce halves that are mirror images of each other?

Introduce the problem. Then have students do the activity to solve the problem. Distribute pattern blocks, GeoReflector™ mirrors, paper, and pencils to students.

Materials
- pattern blocks (1 hexagon and 7 trapezoids per group)
- GeoReflector™ mirror (1 per group)
- paper (1 sheet per group)
- pencils (1 per group)

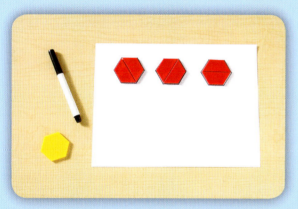

1. Have students use a hexagon block to trace three hexagons onto a sheet of paper, all oriented the same way. **Say:** *Use trapezoid blocks to fill in the hexagons. Show the different ways a hexagon can be cut in half.* Guide students to orient the "cut" lines three different ways.

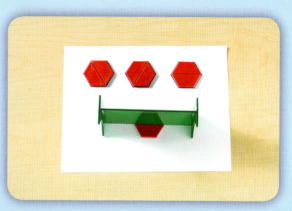

2. Have students select another trapezoid block and use it with the GeoReflector mirror to "construct" a regular hexagon. **Say:** *Half of the hexagon is formed by the trapezoid block and the other half is formed by the image in the mirror.* Introduce the concepts of *symmetry, line of symmetry,* and *mirror image.*

⚠ Look Out!

Students might think that a line of symmetry is any line that divides a shape into two equal halves. Reiterate that the halves must be mirror images of each other in order for the dividing line to be classified as a line of symmetry. Have students investigate this idea with a simple shape, such as a rectangle, and a GeoReflector mirror.

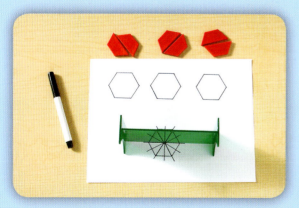

3. Say: *Trace another hexagon onto your sheet of paper. Use the mirror to find and draw all the lines of symmetry of the shape.*

Lesson 6

Geometry

Parallel and Perpendicular Lines

When students understand relationships between lines, they are better equipped to describe shapes and other spatial relationships. Parallel lines are lines in the same plane that do not intersect — that is, do not meet or cross. Perpendicular lines are lines that intersect to form a right angle at the point where they cross.

Try It! Perform the Try It! activity on the next page.

Objective
Identify parallel and perpendicular lines.

Skills
- Locating points on a coordinate grid
- Using spatial relationships
- Recognizing properties of polygons

NCTM Expectations

Grades 6–8
Geometry
- Use coordinate geometry to represent and examine the properties of geometric shapes.
- Use coordinate geometry to examine special geometric shapes, such as regular polygons or those with pairs of parallel or perpendicular sides.
- Recognize and apply geometric ideas and relationships in areas outside the mathematics classroom, such as art, science, and everyday life.

Talk About It

Discuss the Try It! activity.

- **Ask:** *How can you tell if lines are perpendicular to each other?*
- **Ask:** *If a line is parallel to the x-axis, what is its relationship to the y-axis?*
- Have students identify parallel lines and perpendicular lines in the classroom.
- **Say:** *Name some other polygons that have parallel and perpendicular lines.*

Solve It

Reread the problem with students. Have students draw a map of the park and show the paths. Then have them describe the layout of the paths using the terms *parallel* and *perpendicular*.

More Ideas

For other ways to teach about parallel and perpendicular lines—

- Have students use AngLegs™ to create polygons. Have them identify parallel and perpendicular sides.
- Create other polygons on the geoboard and have students describe the relationships of the lines to one another.

Standardized Practice

Have students try the following problem.

Which segments are perpendicular?

A. $\overline{AB}$ and $\overline{AC}$
B. $\overline{AB}$ and $\overline{BC}$
C. $\overline{BC}$ and $\overline{AC}$
D. None of these

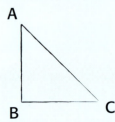

Try It! 20 minutes | Groups of 4

Here is a problem about identifying parallel and perpendicular lines.

Cassidy is landscaping a new city park on a square plot of land. She makes a map of the park to decide where the main paths should lie. She decides to put paths around the perimeter and through the center of the park. How can she describe the relationships of the paths to one another?

Introduce the problem. Then have students do the activity to solve the problem. Distribute geoboards, rubber bands, tape, paper, and pencils to students. **Say:** *The geoboard represents the park and rubber bands will represent the paths.*

Materials
- geoboards (1 per group)
- rubber bands (6 per group)
- masking tape (1 roll per group)
- paper (1 sheet per group)
- pencils (1 per group)

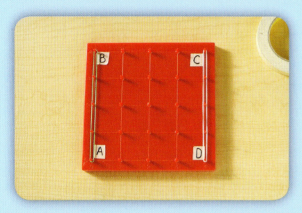

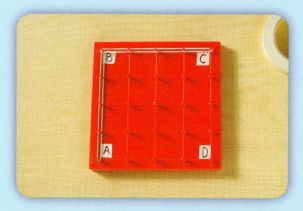

1. Have students locate points at (0, 0), (0, 4), (4, 4), and (4, 0) on the geoboard and label them A, B, C, D respectively. **Say:** *Use rubber bands to connect points A and B and points C and D.* Have students compare the segments by noticing the distance between the line segments. Define *parallel lines*. Have students show other segments that are parallel to $\overline{AB}$.

2. Have students show segments between points A and B and points B and C. **Say:** *Compare these two segments. Define perpendicular lines.* Have students model other perpendicular segments.

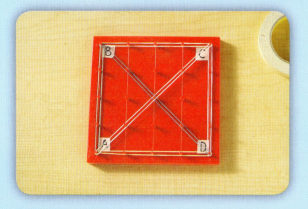

⚠ Look Out!

Have students who confuse the terms *parallel* and *perpendicular* make vocabulary cards and use them to find examples in letters and mathematical symbols. For example, the plus sign has perpendicular lines and the equal sign has parallel lines.

3. Have students connect the four points to form a square. Have them connect the diagonals of the square. **Ask:** *What do you notice about the segments formed by the diagonals?*

Lesson 7

Geometry

Shapes in the Coordinate Plane

Many concepts of geometry can be illustrated on a coordinate grid. Viewing a shape on a coordinate grid helps students identify such properties of the shape as the number and lengths of its sides, the relationships between its sides, and in some cases the measures of its angles. Students who know these properties will more readily understand the formulas for area and perimeter.

Try It! Perform the Try It! activity on the next page.

Objective

Draw shapes on a coordinate grid and describe their properties.

Skills

- Locating points on a coordinate grid
- Using spatial relationships
- Identifying attributes
- Comparing

NCTM Expectations

Grades 6–8
Geometry

- Use coordinate geometry to represent and examine the properties of geometric shapes.
- Use coordinate geometry to examine special geometric shapes, such as regular polygons or those with pairs of parallel or perpendicular sides.
- Draw geometric objects with specified properties, such as side lengths or angle measures.
- Recognize and apply geometric ideas and relationships in areas outside the mathematics classroom, such as art, science, and everyday life.

Talk About It

Discuss the Try It! activity.

- **Ask:** *What do you notice about the coordinates for the square?*
- **Ask:** *How does the coordinate grid help you describe the shapes?*
- Compare the square and the rhombus.
- Compare the square and the triangle.

Solve It

Reread the problem with students. Have students talk about the arrangement of the furniture using coordinates. Have them talk about how the grid helps Beth describe the placement and shape of each piece of furniture.

More Ideas

For other ways to teach about shapes in the coordinate plane—

- Have students make polygons on a geoboard and describe the shapes using coordinates. Ask them to compare the lengths of the sides and the angles and the relationships between the sides of each figure as they describe it.
- Use AngLegs™ and a 4-Quadrant Graph (BLM 14) to make squares, rhombuses, rectangles, parallelograms, and triangles. Tell students to move the shapes around the four-quadrants and name the coordinates of each new position.

Standardized Practice

Have students try the following problem.

If the shape is moved up two units, what coordinates describe the shape's points after the move?

A. Rhombus; (-2, 2), (-2, -2), (2, 2) and (2, -2)

B. Rhombus; (-4, 2), (-4, -2), (6, 2) and (6, -2)

C. Rectangle; (-4, 4), (-4, 0), (4, 4) and (4, 0)

D. Rectangle; (-2, 2), (-2, -2), (6, 2) and (6, -2)

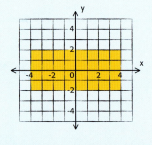

Try It! 30 minutes | Groups of 4

Here is a problem about drawing and describing shapes in the coordinate plane.

Beth is making a record of how the furniture on her patio is positioned. She divides the patio into quadrants and uses a 4-quadrant grid to show the placement of the furniture. She has a square chair, a triangular stool, and a table shaped like a rhombus. Show how the positions and shapes of the pieces of furniture might be represented on the grid.

Introduce the problem. Then have students do the activity to solve the problem. Distribute AngLegs™, graph paper, and pencils to students. Explain to students that the endpoints of the AngLegs™, are represented by the raised circles, not the extreme ends.

Materials
- AngLegs™ (orange and purple only: one set per group)
- 4-Quadrant Graph Paper (BLM 14: 1 per group)
- pencils (1 per group)

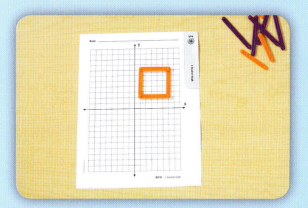

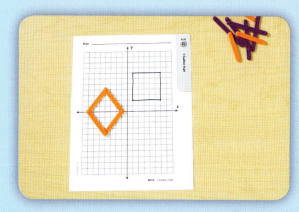

1. Have students use an AngLeg to show a segment connecting points (1, 2) and (1, 7). **Say:** *This segment is one side of the square chair. Complete the chair so that it is positioned completely inside the first quadrant. Draw the shape.* Discuss with students how they found the additional coordinates.

2. Say: *One vertex of the rhombus-shaped table is at point (-4, 4) and another is at (-4, -4). Have students connect the sides to make the rhombus.* **Ask:** *What are the other coordinates of the rhombus?*

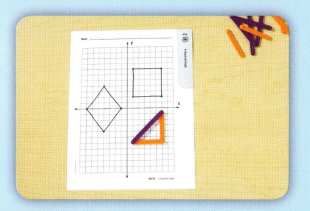

⚠ Look Out!

Some students might have difficulty determining how to connect the points correctly. If students use diagonals to stabilize the rhombus and square, remind them that only the sides of each shape are supposed to be represented; not the diagonals.

3. Have students use an AngLeg to show a segment connecting points (6, -1) and (6, -6). **Say:** *This segment is one side of the stool. Complete the stool so that there is a right angle at (6, -6).*

Lesson 8

Geometry

Slides and Flips

Students develop their understanding of slides and flips by learning how to use coordinates to identify and describe transformations. They learn that slides, or translations, do not change the orientation of a figure, while flips, or reflections, do. These concepts prepare students for the study of multiple transformations.

Objective

Identify and describe slides and flips.

Skills

- Describing shapes
- Identifying quadrants
- Plotting points on a coordinate grid

NCTM Expectations

Grades 6–8
Geometry
- Use coordinate geometry to represent and examine the properties of geometric shapes.
- Describe sizes, positions, and orientations of shapes under informal transformations such as flips, turns, slides, and scaling.

Try It! Perform the Try It! activity on the next page.

Talk About It

Discuss the Try It! activity.

- **Ask:** *In which direction do you slide the triangle along the x-axis?* Have students describe the transformation.
- **Ask:** *Does the size of the triangle change when you slide it to a new position? Does the orientation change?*
- **Ask:** *When you flip the triangle over the x- or y-axis, does the distance of each vertex from the x- or y-axis change or stay the same?*
- **Ask:** *Does the orientation of the triangle change when you flip it over the x- or y-axis? Explain.*

Solve It

Reread the problem with students. Have them write a short paragraph explaining how they can use coordinates to identify and describe a slide or a flip. Have them use the coordinates in their transformations as examples.

More Ideas

For other ways to teach about slides and flips—

- Have students work in pairs. One student uses centimeter cubes to plot corners and draw a figure on a 4-Quadrant Graph (BLM 14). The student labels the vertices with letters and gives instructions to either slide the figure a certain number of units in a specified direction or flip it over one of the axes. The second student performs the transformation, using a GeoReflector™ mirror if needed, and draws and labels the new figure.

- Have students divide a geoboard into 4-quadrants using rubber bands for the x- and y- axes. Tell a student to make a figure and move it using a slide or flip. Have another student identify the type of transformation and give the new coordinates.

Standardized Practice

Have students try the following problem.

What will the new coordinates of point A be if the figure is flipped over the y-axis?

A. (–4, –2)   C. (2, 4)

B. (–2, –4)   D. (4, 2)

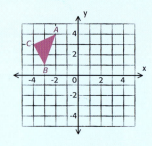

Try It! 30 minutes | Pairs

Here is a problem about slides and flips.

James is using a drawing program on his computer. The program allows him to create and move polygons on a 4-quadrant grid. He is moving a shape using slides and flips. How can he use the coordinates of the shape to identify whether each move was a slide or flip?

Introduce the problem. Then have students do the activity to solve the problem. Distribute AngLegs™, graph paper, GeoReflector™ mirrors, and colored pencils. Explain that a slide moves a figure in a straight line and a flip shows the mirror image of a figure across a line.

Materials
- AngLegs™ (2 orange and 1 purple per pair)
- GeoReflector™ mirrors (1 per pair)
- 4-Quadrant Graph Paper (BLM 14; 1 per pair)
- colored pencils (2 different colors per pair)

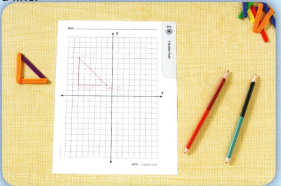

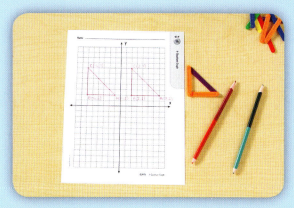

1. Say: *Make a triangle using AngLegs and draw it in Quadrant II with a colored pencil. Label each vertex with a letter and its coordinates using the same color.* Students can label one vertex of the triangle A and use it as a guide for moving the shape to each new location.

2. Say: *Slide the shape along the x-axis to a location in Quadrant I. With the same color as you used for your first drawing, draw the shape and label the points and new coordinates. Label the point that A moved to as A' (read A prime), and so on.* **Ask:** *How have the x- and y-coordinates changed?*

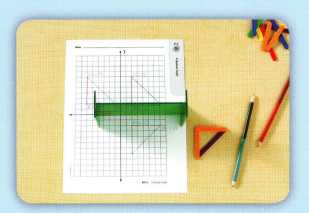

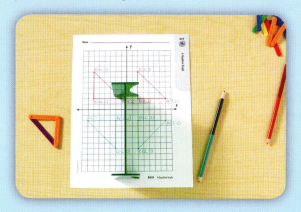

3. Say: *Place the GeoReflector mirror on the x-axis to make an image of the triangle appear in Quadrant IV.* Explain that the image is the shape reflected, or flipped, across the x-axis. **Say:** *Draw and label the flipped shape with the second colored pencil. Label the point that A' moved to A" (read A double prime), and so on.*

4. Say: *Now use the mirror to flip the shape over the y-axis. Draw and label the flipped shape with the second colored pencil. Label the point that A" moved to A'" (read A triple prime), and so on.* **Ask:** *How do the coordinates change when the triangle is flipped across an axis?*

Lesson 9

Geometry
Rotational Symmetry

Students extend their understanding of transformations and symmetry to include rotational symmetry. They use physical models to rotate shapes and recognize when a shape rotates onto itself. An understanding of rotational symmetry strengthens a student's ability to describe and classify shapes.

Objective
Identify whether figures have rotational symmetry.

Skills
- Describing shapes
- Using spatial reasoning
- Using transformations

NCTM Expectations
Grades 6–8
Geometry
- Describe sizes, positions, and orientations of shapes under informal transformations such as flips, turns, slides, and scaling.
- Examine the congruence, similarity, and line or rotational symmetry of objects using transformations.

Try It! Perform the Try It! activity on the next page.

Talk About It
Discuss the Try It! activity.

- **Ask:** *How can you tell that the triangle will not rotate onto itself at $\frac{1}{4}$, $\frac{1}{2}$, and $\frac{3}{4}$ turns?*
- **Ask:** *What is the relationship between the number of vertices and the number of times a shape could be rotated onto itself in one full turn? Does this relationship apply to the other pattern-block shapes?*
- Have students express the fractional turns in degrees.

Solve It
Reread the problem with students. Have them write about rotational symmetry and how it applies to the problem about Ella's pins.

More Ideas
For other ways to teach about rotational symmetry—
- Have students continue the activity with the other pattern blocks.
- Have students do the activity using shapes made with AngLegs™. Have them include both regular and irregular polygons.

Standardized Practice
Have students try the following problem.

Which figure has $\frac{1}{3}$-turn rotational symmetry?

A.   B.   C.   D.

Try It! 30 minutes | Pairs

Here is a problem about rotational symmetry.

Ella has a collection of pins. Her three favorite pins are shaped like a triangle, a square, and a hexagon. She wears each pin a certain way. For example, she likes one of the points of her triangle-shaped pin to be at the top. For each pin, determine how many ways Ella can orient the pin so that it will look the way she likes it.

Introduce the problem. Then have students do the activity to solve the problem. Distribute pattern blocks, tape, paper, and pencils to students. **Say:** *A figure that has rotational symmetry can be rotated onto itself in less than a full turn.*

Materials
- pattern blocks (one of each shape per pair)
- paper (1 sheet per pair)
- pencils (1 per pair)
- masking tape (1 small length per pair)

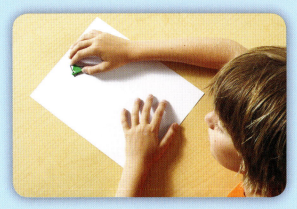

1. Say: *Place the triangle on a sheet of paper with a vertex at the top. Mark this vertex with a piece of tape. Trace the triangle.* Have students rotate the triangle until it is rotated onto itself. **Ask:** *How far did you rotate the triangle until it rotated onto itself? Will the triangle rotate onto itself again if you continue to rotate it?* Guide them to see that it will rotate onto itself at $\frac{1}{3}$ and $\frac{2}{3}$ turns.

2. Have students mark a corner on the square block and trace the block on the sheet of paper. **Ask:** *Will the square rotate onto itself in less than a full turn?* Have students rotate the block. Guide them to see that it will rotate onto itself at $\frac{1}{4}$, $\frac{1}{2}$, and $\frac{3}{4}$ turns.

⚠ Look Out!

Watch for students who do not understand the concept of a figure rotating onto itself in less than a full turn. Reinforce the concept using additional shapes, including shapes that do not have rotational symmetry.

3. Say: *Now use the hexagon pattern block to determine if a regular hexagon will rotate onto itself.* Guide students to see that a hexagon will rotate onto itself at $\frac{1}{6}$, $\frac{2}{6}$, $\frac{3}{6}$, $\frac{4}{6}$, and $\frac{5}{6}$ turns.

Lesson 10

Geometry
Multiple Transformations

After students have learned to perform and identify single slides, flips, and turns, they can move on to multiple transformations, or transformations that are performed one after another on the same figure. In this activity, students see that the result of multiple transformations can be the same as the result of a single transformation.

Try It! *Perform the Try It! activity on the next page.*

Objective
Perform multiple transformations on geometric figures.

Skills
- Reflecting figures
- Identifying transformations
- Using spatial reasoning

NCTM Expectations
Grades 6–8
Geometry
- Describe sizes, positions, and orientations of shapes under informal transformations such as flips, turns, slides, and scaling.

Talk About It
Discuss the Try It! activity.

- **Say:** *When you flip a figure, what characteristics of the figure change and what stays the same?*
- **Say:** *Name the two transformations in Step 2.*
- **Say:** *Name the one transformation in Step 3.*
- **Ask:** *How do you know that the two transformations in Step 2 give the same result as the one transformation? Do you think this always works?*

Solve It
Reread the problem with students. After performing two flips on Lee's stationery, the "L" appears upside down in the bottom right-hand corner. The "L" is in this same position after one turn of 180° about the origin. Lee can use either transformation to load the stationery into the printer.

More Ideas
For other ways to teach about multiple transformations—

- Have students use a GeoReflector™ mirror and figures formed by AngLegs™ to experiment with multiple reflections in parallel lines. Start by drawing two parallel lines, *m* and *n*. Students will see that when a figure is reflected in line *m*, and that image is reflected in line *n*, the result is the same as a single translation (slide) of the original figure. The position of the figure is changed, but the orientation is not changed.

- Students can perform multiple transformations (any combination of slides, flips, and turns) on pattern blocks. A student traces the original figure and the final image, then a partner figures out the transformations.

Standardized Practice
Have students try the following problem.

Which single transformation gives the same result as two reflections if the lines of reflection intersect?

A. slide   **B.** flip   **C.** turn   **D.** shrink

Try It! 20 minutes | Groups of 4

Here is a problem about multiple transformations.

Lee has stationery personalized with her initial "L" at the top left corner. To print a letter from her computer, she needs to feed the stationery into her printer bottom first, with the print facing up. How would you write a rule based on flips and/or turns to help Lee determine how to feed the paper?

Introduce the problem. Then have students do the activity to solve the problem. Distribute GeoReflector™ mirrors, AngLegs™, graph paper, paper, and pencils to students.

Materials
- GeoReflector™ mirrors (1 per group)
- AngLegs™ (3 purple and 3 orange per group)
- 4-Quadrant Graph Paper (BLM 14; 1 per group)
- paper (1 sheet per group)
- pencils (1 per group)

1. Say: *Using AngLegs, build three identical figures shaped like the letter "L". Write an L at the top left corner of a sheet of paper. This is like one page of Lee's stationery.*

2. Say: *Place one of the L models on the graph, in Quadrant II. Place the mirror along the y-axis; then place another L on top of the reflected image. Next, use the mirror to reflect this L across the x-axis, and place your third L on this reflected image.* Students place the L-shaped figures as shown. **Say:** *Now perform these transformations with the stationery.* Students flip the paper over once to the right and then again vertically so the top of the paper becomes the bottom.

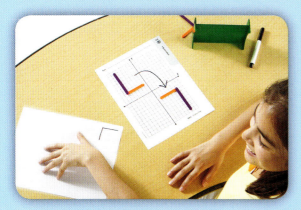

3. Say: *Remove the mirror and the L you placed in Quadrant I.* **Ask:** *What single transformation can you perform on the upper left L to get the lower right L? Show this on your paper.* Students draw an arrow to show a rotation. **Say:** *Perform this transformation with the stationery.* Students return the stationery to the original position and then turn it about the origin 180°.

⚠ Look Out!

Point out that if the stationery is flipped once, rather than twice, the type will print on the back side of the stationery. It must be flipped again to print on the front. Students may not think that two reflections can give the same result as one rotation. Have them try this as many times as necessary to become convinced!

Lesson 11

Geometry

Tessellations

A tessellation is a fascinating combination of geometry and art. It is an arrangement of shapes that cover a surface without gaps or overlapping. Tessellations can be fun to create and beautiful to look at. They can be useful, too. Tessellations can also be analyzed mathematically, which is what students do in this activity.

Objective
Investigate the characteristics of tessellations.

Skills
- Combining shapes
- Measuring angles
- Reasoning

NCTM Expectations

Grades 3–5
Geometry
- Investigate, describe, and reason about the results of subdividing, combining, and transforming shapes.
- Make and test conjectures about geometric properties and relationships and develop logical arguments to justify conclusions.

Try It! Perform the Try It! activity on the next page.

Talk About It

Discuss the Try It! activity.

- **Ask:** *What is true about the sum of the angle measures at each vertex? Are you surprised by this answer?*
- **Say:** *With your figure, point to a vertex in any of your tessellations or designs. Circle around the vertex with your finger.* **Ask:** *How many degrees are there in a full circle?*
- **Ask:** *What is the fewest number of shapes you can use to form a vertex with pattern blocks?*
- **Ask:** *What is the greatest number of shapes you can use to form a vertex with pattern blocks?*

Solve It

Reread the problem with students. The sum of the angle measures at each vertex is 360°. Note that when one shape is used repeatedly, there may be more than one way to combine angles at a vertex. For example, with a blue rhombus, a vertex can be formed by 3 shapes (120° + 120° + 120°), 4 shapes (60° + 120° + 60° + 120°), 5 shapes (120° + 60° + 60° + 60° + 60°), or 6 shapes (60° + 60° + 60° + 60° + 60° + 60°). Other shapes, such as squares, can form a vertex in only one way.

More Ideas

For other ways to teach about tessellations—

- Have students use pattern blocks to create various types of designs: (1) designs with rotational symmetry that start with a yellow hexagon in the center and continue outward to form a *circular* shape; (2) designs with horizontal or vertical line symmetry; and (3) designs with no symmetry.
- Have students work in pairs to create a tessellation. One student begins by choosing any pattern block. Then the other adds a block. Students take turns placing blocks. Challenge them to create a pattern with symmetry.

Standardized Practice

Have students try the following problem.

What is the measure of angle X?

A. 30°   B. 60°   C. 90°   D. 120°

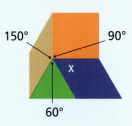

Try It! 30 minutes | Groups of 3 to 6

Here is a problem about tessellations.

Matt is gluing pattern blocks to a board to make a display for his room. He places the blocks in a pattern so there are no gaps or overlaps. What must be true about the sum of the angle measures at each vertex in the pattern?

Introduce the problem. Then have students do the activity to solve the problem. Distribute pattern blocks, protractors, recording charts, and pencils to students. Discuss the characteristics of a tessellation and make sure students understand the term *vertex*.

Materials
- pattern blocks (at least 6 of each color per group)
- protractors (1 per group)
- 4-Column Recording Chart (BLM 17; 2 copies per group)
- pencils (1 per group)

1. Have students create a tessellation for each shape. This works with all the colors of pattern blocks. **Say:** *Make sure you form at least one vertex for each tessellation.* Students use at least six of each of the following: yellow hexagon, red trapezoid, blue rhombus, orange square, green triangle, and tan rhombus.

2. Say: *Measure the interior angles for each type of pattern block.* Have students start a table with these columns: *Shape, Sketch, Sum (of Angle Measures at Vertex).* **Ask:** *What is the sum of the angle measures at each vertex?* The sum is always 360°.

⚠ Look Out!

If students get a sum that is not 360°, have them check for errors in measurement or arithmetic. Encourage students to use their table showing the angle measures of each pattern block. For example, when a blue rhombus is part of a vertex, the angle measure is either 60° or 120°. Guide students to choose the correct measure.

3. Ask: *Do you think the sum will be 360° if you use any combination of shapes?* **Say:** *Experiment with this question by creating a design that uses all six colors of pattern blocks. Find the sum of the angle measures at each vertex.* Students can use 3 to 12 shapes to form a vertex. The sum is always 360°.

Lesson 12

Geometry

Congruent Figures and Transformations

Congruent figures are exactly the same size and shape. Manipulatives help students visualize this. They can match figures and then move the figures to different positions and orientations, noting that the model does not change. They can also reverse the process to determine if figures in different positions are congruent.

Try It! Perform the Try It! activity on the next page.

Objective

Identify congruent figures.

Skills

- Identifying attributes
- Comparing
- Applying transformations

NCTM Expectations

Grades 3–5
Geometry
- Identify, compare, and analyze attributes of two- and three-dimensional shapes and develop vocabulary to describe the attributes.
- Explore congruence and similarity.
- Describe a motion or a series of motions that will show that two shapes are congruent.
- Recognize geometric ideas and relationships and apply them to other disciplines and to problems that arise in the classroom or in everyday life.

Talk About It

Discuss the Try It! activity.

- **Ask:** *How can you tell if two figures are congruent?*
- **Ask:** *Are figures still congruent when they are moved in more than one way?* Have students show two transformations, such as a reflection, and a rotation, to verify their responses.
- Have students identify congruent figures in the classroom and recognize how different positions could be the result of translations, reflections, rotations, or combinations of transformations.

Solve It

Reread the problem with students. Have them draw a design for Calvin's window using four congruent rectangles. Then have them explain how to tell that the figures are congruent.

More Ideas

For other ways to teach about congruency and transformations—

- Use pattern blocks to show congruency. Have students use different shapes to make congruent hexagons. Have them move the shapes to show that position and orientation do not affect congruency.
- Adapt the problem to use a different shape, such as a triangle. Have students use AngLegs™ to solve the problem.

Standardized Practice

Have students try the following problem.

Which shape is congruent to this rectangle?

A.   B.   C.   D.

Try It! 25 minutes | Groups of 4

Here is a problem about congruent figures and transformations.

Calvin makes stained-glass windows. He lays out 4 rectangles to be used in one of his designs. The rectangles are arranged in two rows and two columns. How can he move the rectangles to determine if they are congruent?

Introduce the problem. Then have students do the activity to solve the problem. Distribute AngLegs™ to students. **Say:** *Remember there are three types of transformations: translations or slides, reflections or flips, and rotations or turns.* Draw the arrangement of the rectangles on the board.

Materials
- AngLegs™ (1 set per group)

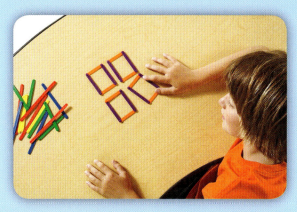

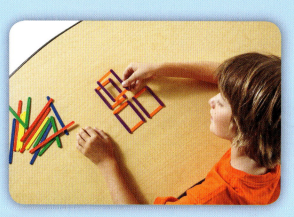

1. Have students use AngLegs to make a rectangle. **Say:** *Make three more rectangles that are the same size and shape as the one you just made.* Have students arrange the four rectangles as shown on the board.

2. Have students slide the top right-hand model to the left on top of the other rectangle. **Say:** *Compare the two rectangles. Are they the same size and shape?*

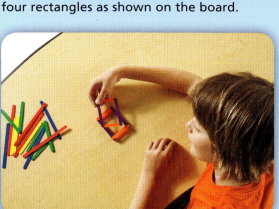

⚠ Look Out!

Some students may have difficulty maintaining the shape of the AngLegs when moving the models. You might need to have them brace the figure by connecting one or more diagonals. Remind them that the brace does not change the characteristics of the larger shape. Students who think that position and orientation affect congruency may need more hands-on practice with a variety of shapes.

3. Have students reflect the bottom left rectangle onto the top rectangle. Then have them rotate the last rectangle onto the top of the stack of rectangles. Guide students to see that the size and shape of each rectangle remains the same when the position and/or orientation is changed. Explain that the figures are *congruent* because they have exactly the same size and shape.

Lesson 13

Geometry

Corresponding Parts of Congruent Figures

Corresponding parts of congruent figures match exactly and have equal measures. Recognizing these relationships in triangles allows students to informally explore side and angle relationships used in geometric proofs. It also prepares students to identify congruent angles in similar figures and to use proportionality to find the side lengths in similar figures.

Try It! Perform the Try It! activity on the next page.

Objective
Identify corresponding parts of congruent figures.

Skills
- Identifying attributes
- Comparing
- Reasoning

NCTM Expectations

Grades 6–8
Geometry
- Precisely describe, classify, and understand relationships among types of two- and three-dimensional objects using their defining properties.
- Create and critique inductive and deductive arguments concerning geometric ideas and relationships, such as congruence, similarity, and the Pythagorean relationship.
- Describe sizes, positions, and orientations of shapes under informal transformations such as flips, turns, slides, and scaling.
- Recognize and apply geometric ideas and relationships in areas outside the mathematics classroom, such as art, science, and everyday life.

Talk About It

Discuss the Try It! activity.

- **Ask:** *How can you identify corresponding parts in congruent triangles when the triangles are in different positions?*
- Draw two congruent isosceles triangles on the board. Have students identify corresponding sides and corresponding angles.
- Discuss that corresponding sides and angles in congruent figures always have the same measure.

Solve It

Reread the problem with students. Have students draw two congruent frames and use different colors to mark the pairs of corresponding sides and corresponding angles in the two figures. Ask them to explain how Joshua can use corresponding parts to make sure the frames are congruent.

More Ideas

For other ways to teach about corresponding parts of congruent figures—

- Use geoboards to make congruent figures. Have students use same-color bands to show corresponding sides. They can visually compare corresponding angles by referring to the color of the segments making up the angles.
- Have students use AngLegs™ to model congruency in different polygons.

Standardized Practice

Have students try the following problem.

Which is a pair of corresponding parts in these congruent triangles?

A. ∠A and ∠B
B. ∠D and ∠B
C. $\overline{AC}$ and $\overline{DF}$
D. $\overline{CB}$ and $\overline{DE}$

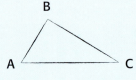
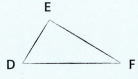

Try It! 30 minutes | Groups of 4

Here is a problem about corresponding parts of congruent figures.

Joshua is building triangular wood frames to construct a ramp. How can he make sure that each triangular frame is congruent with the other frames?

Introduce the problem. Then have students do the activity to solve the problem. Distribute AngLegs™, paper and colored pencils to students. Review the attributes of a right triangle.

Materials
- AngLegs™ (1 set per group)
- paper (1 sheet per group)
- colored pencils (several per group)

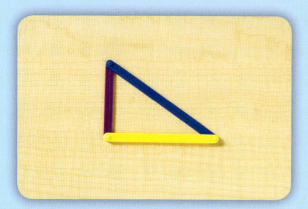

1. Have students use one purple, one yellow, and one blue AngLeg to make a right triangle.

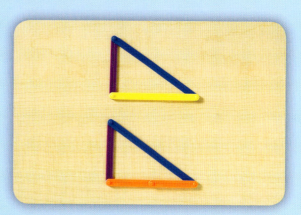

2. Say: *Use two orange, one blue, and one purple AngLeg to make another triangle that is the same size and shape.* Elicit that the two triangles are congruent.

 Look Out!

Students who have difficulty seeing corresponding sides and angles may need to snap the figures together to show congruency. They can pick up the two figures and physically touch the corresponding parts. When the figures have different orientations, have students match colors when possible.

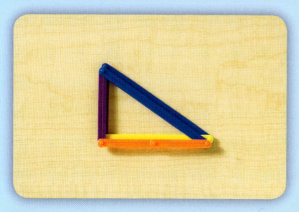

3. Have students stack the triangles so the sides exactly match. **Ask:** *Which side exactly matches the yellow side?* Guide students to see that corresponding sides in congruent figures have the same length and have the same relationship to the other sides in the figure. **Ask:** *Do the angles formed by those corresponding sides also have the same measure? How do you know?*

Lesson 14

Geometry
Similar Triangles

When students explore similarity, they first learn that a figure can be a smaller or larger version of another figure of the same shape. Students expand their understanding of similarity as they study proportions, when they learn that corresponding sides of similar figures are proportional.

Try It! Perform the Try It! activity on the next page.

Objective
Build similar triangles.

Skills
- Identifying isosceles triangles
- Recognizing similar triangles
- Measuring angles

NCTM Expectations
Grades 3–5
Geometry
- Explore congruence and similarity.
- Build and draw geometric objects.
- Recognize geometric ideas and relationships and apply them to other disciplines and to problems that arise in the classroom or in everyday life.

Talk About It
Discuss the Try It! activity.

- **Say:** *Similar triangles have the same shape but not necessarily the same size. Are the angles of the blue triangle congruent to the corresponding angles of the green triangle? Are corresponding sides of the two triangles congruent?*
- **Say:** *Look at your four isosceles right triangles.* **Ask:** *Why are these triangles similar to one another but not to the equilateral triangles?*
- **Say:** *Look at your two triangles that have a 120° vertex angle. Find the side lengths written on the AngLegs™. Compare the ratio blue/green with the ratio purple/orange. What do you find?*
- **Say:** *Snap the six equilateral triangles together at one of their vertices. What do you notice? Do the same for the 4 right triangles. What do you notice?* Help students see that the sides opposite the angles that are snapped together are parallel to one another.

Solve It
Reread the problem with students. Have students explain what *similar* means and how the concept relates to the problem.

More Ideas
For other ways to teach about similar triangles—

- Students can use AngLegs to build more similar triangles. For example:
  (R = red, B = blue, Y = yellow, G = green, P = purple, O = orange).
  Isosceles triangles: RRB and YYG, BBR and GGY, YYB and PPG, BBY and GGP, GGO and BBP, RRY and BBG and PPO and YYP.
  Scalene triangles: BPR and GOY (30°–60°–90°), POG and YPB (right), POY and YPR (obtuse), BYR and GPY (acute).
- Ask students to place any triangle formed by AngLegs on a sheet of paper. Trace around the outside, then trace around the inside of the legs. Explain why the two triangles they just drew are similar.

Standardized Practice
Have students try the following problem.

Which of the following is a pair of similar triangles?

A.   B.   C.   D.

86

Try It! 30 minutes | Groups of 6

Here is a problem about similar triangles

Will is drawing a doghouse in various sizes for a cartoon. He is experimenting with isosceles triangles for the roof design. Consider some possible roof shapes and show how a particular shape can be made in different sizes.

Introduce the problem. Then have students do the activity to solve the problem. Distribute AngLegs™ and protractors to students.

Materials
- AngLegs™ (1 set per group)
- protractors (1 per group)

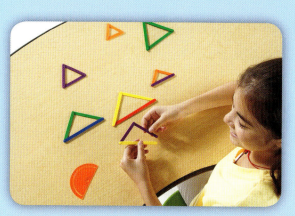

1. Say: *Build 6 different sizes of equilateral triangles.* Students build 6 equilateral triangles, each formed by 3 same-color legs. **Ask:** *What is the measure of each angle?* Each angle of every triangle measures 60°. Explain that these triangles are similar to each other because they are the same shape.

2. Say: *Now build 4 different sizes of isosceles triangles having a 90° vertex angle. These are isosceles right triangles, or 45-45-90 triangles for their angle measures.* Students build 4 triangles using 2 yellow and 1 red, 2 green and 1 blue, 2 purple and 1 yellow, and 2 orange and 1 purple. **Say:** *These triangles are similar to each other, but not to the ones you built first.*

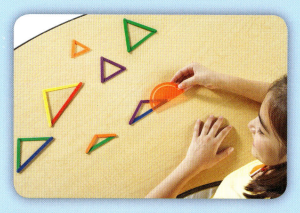

⚠ Look Out!

Because the word similar is used in everyday language, students might not know that it has a mathematical definition as well. They might think, for example, that all triangles are similar because they all have three sides. But in order for triangles to be similar in the mathematical sense, they must have the same shape. This means that all corresponding angles must be congruent and all corresponding sides must be proportional.

3. Say: *Next, build 2 different sizes of isosceles triangles having a 120° vertex angle. Use the protractor to check the angle measures.* Students use 2 purple and 1 blue and 2 orange and 1 green. **Say:** *These triangles are similar to each other, but not to the ones you built earlier. All of the sets are possible shapes for the doghouse roof.*

Lesson 15

Geometry

Nets

By constructing three-dimensional figures from two-dimensional representations, students see how plane shapes can be related to solid shapes. Identifying the faces, edges, and vertices on both nets and solid polyhedrons helps students develop the visualization skills needed to find surface area.

Try It! Perform the Try It! activity on the next page.

Objective
Explore nets.

Skills
- Constructing three-dimensional shapes
- Comparing two- and three-dimensional shapes
- Using spatial visualization

NCTM Expectations

Grades 3–5
Geometry
- Identify, compare, and analyze attributes of two- and three-dimensional shapes and develop vocabulary to describe the attributes.
- Investigate, describe, and reason about the results of subdividing, combining, and transforming shapes.
- Build and draw geometric objects.
- Identify and build a three-dimensional object from two-dimensional representations of that object.

Talk About It
Discuss the Try It! activity.
- **Ask:** *How do you identify the faces in a net?*
- **Ask:** *How do you make a net for a rectangular box?*

Solve It
Reread the problem with students. Display a rectangular prism and explain that it represents a rectangular box. Have students sketch the solid and draw its net. Then have them write a paragraph explaining how Cameron can tell which cardboard piece is a net for a rectangular box.

More Ideas
For other ways to teach about nets—

- Have students use Relational GeoSolids® to make more nets. Guide them to trace the base; then, without lifting the solid, turn the figure and trace a different face. Be sure they understand that the faces share edges but do not overlap. Have students cut out the nets and construct the solids.

- Show students the GeoSolid cube. Explain what a net is. Point out the faces, edges, and vertices on a net. Challenge students to make 10 different nets for a cube.

Standardized Practice
Have students try the following problem.

What net can be folded to make this solid?

A.   B.

C.   D.

Try It! 30 minutes | Groups of 4

Here is a problem about nets.

Cameron works for a shipping company. Each shipping carton is stored as a flat cardboard piece that can be assembled when needed. What might the cardboard piece for a rectangular box look like?

Introduce the problem. Then have students do the activity to solve the problem. Distribute solids, Net Patterns, paper, pencils, scissors, and tape to students. Explain that a net is a pattern that can be folded to make a three-dimensional object.

Materials
- Relational GeoSolids® (one set per group or two groups can share a set)
- Net Pattern (BLM 16; 1 per group)
- paper (2 sheets per group)
- pencil (1 per group)
- scissors (1 per group)
- tape (1 per group)

1. Have students examine the solids and describe how the shapes differ from two-dimensional shapes, such as rectangles. Display a cube and point out a face, an edge, and a vertex, defining each. **Say:** *Find a solid that has 12 vertices.*

2. Have students look at the net pattern and identify the faces and their shapes, the edges, and the vertices in the net. **Ask:** *Which solid figure has faces that are the same shape as the shapes in the net?* Have students cut out the net and fold it to make a solid.

⚠ Look Out!

Some students may be inclined to cut out the faces and tape them together to form a solid. Point out that while this method can be used to create a three-dimensional shape, a net is a single piece that can be folded to make the shape. Instruct students to cut only along the outside of the plane figure.

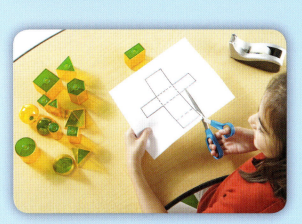

3. Have students use one of the two rectangular prisms to make a net for a rectangular box. **Say:** *Cut and build your rectangular box from your net.*

Lesson 16

Geometry
Three-Dimensional Shapes

Students must be able to identify and characterize faces, edges, and vertices in order to classify three-dimensional shapes. They generalize the relationship between faces, edges, and vertices using Euler's Formula: $V + F - E = 2$, where V represents the number of vertices, F represents the number of faces, and E represents the number of edges.

Try It! Perform the Try It! activity on the next page.

Objective
Classify three-dimensional shapes.

Skills
- Classifying shapes
- Analyzing attributes
- Comparing

NCTM Expectations

Grades 6–8
Geometry
- Precisely describe, classify, and understand relationships among types of two- and three-dimensional objects using their defining properties.
- Recognize and apply geometric ideas and relationships in areas outside the mathematics classroom, such as art, science, and everyday life.

Talk About It
Discuss the Try It! activity.

- **Ask:** *What is opposite a base in a prism? What is opposite the base in a pyramid?*
- **Ask:** *What shape has only triangles for faces? What shape has only squares for faces?*
- **Ask:** *Why is a cube also a rectangular prism but a rectangular prism may or may not be a cube?*

Solve It
Reread the problem with students. Have students write a paragraph to describe two ways to arrange the candles when displaying them by shape. Have students explain which attributes they used to classify the shapes. Guide students to determine the relationship between number of vertices, number of faces, and number of edges.

More Ideas
For other ways to teach about classifying three-dimensional shapes—

- State properties of a specific three-dimensional figure and have students find a solid in the set of Relational GeoSolids® that has those properties. Some statements, such as "I have two rectangular-shaped bases." may be true for multiple figures. Others may apply to only one solid: "My only base is a circle."

- Display a classroom object. Have students describe its attributes, and then find which Relational GeoSolid has the same attributes.

Standardized Practice
Have students try the following problem.

What three-dimensional shape is made up of four triangles?

A. triangle

B. triangular prism

C. triangular pyramid

D. hemisphere

Try It! 30 minutes | Groups of 4

Here is a problem about classifying three-dimensional shapes.

The Candle Shop changes its displays often. What characteristics might Juliana consider if she wants to arrange the candles in a display by shape? Juliana has discovered a relationship among the number of faces, vertices, and edges of the polyhedron-shaped candles. What is it?

Introduce the problem. Then have students do the activity to solve the problem. Distribute the solids, paper, and pencils to students. Review the names of the solids. Point out that the base of each figure is the green face. Remind students that a polygon is a closed two-dimensional shape with sides that are straight lines.

Materials
- Relational GeoSolids® (1 set per group)
- paper (1 sheet per group)
- pencils (1 per group)

1. Have students select all the solids that have a base that is a polygon. **Say:** *These shapes are called polyhedrons.* Have students start a chart that has the following headings: Polyhedron Name, Shape of Base(s), Shape of Other Faces, Number of Faces, Number of Vertices, Number of Edges.

2. Have students find polyhedrons that have opposite bases that are parallel and congruent. **Say:** *These shapes are prisms.* Have students identify the attributes of the other polyhedrons. **Say:** *These shapes are pyramids.* Explain that both prisms and pyramids are named by the shape of their base. **Say:** *Suggest a way that Juliana can organize the candles by shape.*

⚠ Look Out!

Students who confuse the names of prisms and pyramids may need more experience with models of these figures. They might find it helpful to learn about specific real-life examples, such as the Transamerica Building in San Francisco or the pyramids of Egypt. Others may use the first letters of the word **p**rism to think **p**arallel, **r**ight angle to remind themselves of the parallel bases and the angle formed by each base and a nonbase face.

3. Have students complete the chart. *Tell them that there is a relationship between the number of faces, edges, and vertices.* Challenge students to determine the relationship and write it as an equation.

Algebra

Algebra uses symbols to represent mathematical relationships and to solve problems. It builds on students' experiences with numbers and is closely linked to geometry and data analysis. In this way, the ideas taught in algebra help to unify the mathematics program.

In Grades 5 and 6, students are introduced to the use of equations as a means of representing mathematical relationships. Students learn to interpret letter symbols as placeholders for values that have not yet been assigned or determined. They learn that when such placeholders, or variables, appear in an equation, the equation expresses a relationship between the quantities that the placeholders represent. Students understand this by using their sense of pattern. For example, they interpret a two-variable equation as a kind of pattern generator by which each input generates a specific output. The "pattern" resides in the order, or predictability, with which the value of the input determines the value of the output. Students demonstrate the thought process by building function tables.

The NCTM Standards for Algebra suggest that students should:

- Understand patterns, relations, and functions
- Represent and analyze mathematical situations and structures using algebraic symbols
- Use mathematical models to represent and understand quantitative relationships
- Analyze change in various contexts

Students at these grade levels learn how to represent linear relationships using graphs. In doing this, they apply what they are learning about two-variable equations. They generate coordinate pairs, plot them on a coordinate grid, and recognize that the plotted points are a visual representation of the equation. Students also begin working with integers in these grade levels, so they learn about all four quadrants of the rectangular coordinate system and learn how to use all four quadrants to graph linear equations. The following activities are built around manipulatives that students can use to develop skills and explore concepts in **Algebra**.

Algebra
Contents

Lesson 1   Properties of Addition 94
  Objective: Identify and use commutative, associative, and identity properties of addition.
  Manipulative: Cuisenaire® Rods

Lesson 2   Properties of Multiplication 96
  Objective: Identify and use commutative, associative, and identity properties of multiplication.
  Manipulative: Cuisenaire® Rods

Lesson 3   Distributive Property 98
  Objective: Identify and use the Distributive Property.
  Manipulative: base ten blocks

Lesson 4   Order of Operations 100
  Objective: Use the order of operations to simplify expressions.
  Manipulative: color tiles

Lesson 5   Expressions with a Variable 102
  Objective: Write and evaluate an expression with a variable.
  Manipulative: color tles

Lesson 6   Equations with a Variable 104
  Objective: Write and solve an equation with a variable.
  Manipulative: Cuisenaire® Rods

Lesson 7   Addition and Subtraction Equations 106
  Objective: Solve addition and subtraction equations.
  Manipulative: Cuisenaire® Rods

Lesson 8   Multiplication and Division Equations 108
  Objective: Solve multiplication and division equations.
  Manipulative: Cuisenaire® Rods

Lesson 9   Patterns and Function Tables 110
  Objective: Use patterns and function tables to solve problems.
  Manipulative: pattern blocks

Lesson 10   Introduction to Integers 112
  Objective: Represent, compare, and order integers.
  Manipulative: color tiles

Lesson 11   Add Integers 114
  Objective: Add integers.
  Manipulative: two-color counters

Lesson 12   Subtract Integers 116
  Objective: Subtract integers.
  Manipulative: two-color counters

Lesson 13   Multiply Integers 118
  Objective: Multiply integers.
  Manipulatives: two-color counters

Lesson 14   Divide Integers 120
  Objective: Divide Integers.
  Manipulative: two-color counters

Lesson 15   4-Quadrant Graphing 122
  Objective: Graph points in the four quadrants of the coordinate plane.
  Manipulative: centimeter cubes

Lesson 16   Graphing Linear Equations 124
  Objective: Graph linear equations on a four-quadrant grid.
  Manipulatives: centimeter cubes

Lesson 1

Algebra
Properties of Addition

An understanding of the properties of addition helps students to calculate mentally and prepares them for working with equations. The Commutative (or order) Property of Addition states $x + y = y + x$. The Associative (or grouping) Property of Addition states $x + (y + z) = (x + y) + z$. The Identity Property of Addition states $a + 0 = a$.

Objective
Identify and use commutative, associative, and identity properties of addition.

Skills
- Adding
- Representing numbers

NCTM Expectations
Grades 3–5
Algebra
- Identify such properties as commutativity, associativity, and distributivity and use them to compute with whole numbers.
- Express mathematical relationships using equations.
- Model problem situations with objects and use representations such as graphs, tables, and equations to draw conclusions.

Try It! Perform the Try It! activity on the next page.

Talk About It
Discuss the Try It! activity.

- **Ask:** *How do the models for 6 + 3 = 3 + 6 show the Commutative Property of Addition?*
- **Ask:** *How do the parentheses in 6 + (3 + 4) and (6 + 3) + 4 change the way you do the addition?* Talk about the Associative Property of Addition with students.

Solve It
Reread the problem with students. Have students draw each of their models on grid paper. Then have students explain which property is shown by each model.

More Ideas
For other ways to teach properties of addition—

- Use centimeter cubes to model each of the three properties of addition. Have students discover the characteristics of each property and record a copy of the model and the related number sentences for each property on centimeter grid paper.
- Use Snap Cubes® to model various problems. Write number sentences on the board showing each property, such as *4 + 5 = 5 + 4*. Have students compare the lengths in each model to see that they are equal. Explain which property is illustrated by each model.

Standardized Practice
Have students try the following problem.

Which number sentence shows the Associative Property of Addition?

A. 18 + 12 = 12 + 18   **B.** 8 + 0 = 8

C. 6 + 1 = 7   **D.** 14 + (11 + 7) = (1 + 11) + 7

Try It! 30 minutes | Groups of 4

Here is a problem about the properties of addition.

A package contains 6 large glow in-the-dark-balls, 3 small glow-in-the-dark balls, and 4 balls that don't glow in the dark. Kai and Ty use different ways to find the total number of glow-in-the-dark balls, and they use two different ways to find the total number of balls. Show the different ways Kai and Ty add the balls.

Introduce the problem. Then have students do the activity to solve the problem. Review the meaning of parentheses in expressions. Distribute the Cuisenaire® Rods, grid paper, and colored pencils to students.

Materials
- Cuisenaire® Rods (1 set per group)
- Centimeter Grid Paper (BLM 8; 1 per group)
- colored pencils

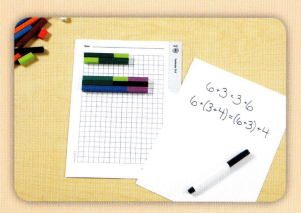

1. Have students show two ways to find the number of glow-in-the-dark balls: add the large balls first, and then add the small balls first. Ask them to compare the two models on the grid paper. Write $6 + 3 = 3 + 6$ on the board. Explain that the models show the Commutative Property of Addition.

2. Write $6 + (3 + 4)$ and $(6 + 3) + 4$ on the board. **Say:** *Kai and Ty use different ways to add all the balls in the package. Model both ways.* Ask students to compare the models and the expressions on the grid paper. Introduce the Associative Property of Addition.

3. Say: *If no more balls are added to the package, the package does not change.* Write $13 + 0$ on the board. Have students model the expression. **Ask:** *What is the sum of zero and any number?* Introduce the Identity Property of Addition.

⚠ Look Out!

Some students may generalize these addition properties to include subtraction. Have them use calculators to discover that subtraction has different results when the order and/or grouping of the numbers is changed. Other students may try to use the Associative Property of Addition to group different operations. Stress that only one operation pertains to the property, as stated in the full name, Associative Property of Addition.

Lesson 2

Algebra
Properties of Multiplication

An understanding of the properties of multiplication helps students to calculate mentally and prepares them for working with equations. The Commutative (or order) Property of Multiplication states $a \times b = b \times a$. The Associative (or grouping) Property of Multiplication states $a \times (b \times c) = (a \times b) \times c$. The Identity Property of Multiplication states $b \times 1 = b$.

Try It! *Perform the Try It! activity on the next page.*

Objective
Identify and use commutative, associative, and identity properties of multiplication.

Skills
- Multiplying
- Representing numbers

NCTM Expectations
Grades 3–5
Algebra
- Identify such properties as commutativity, associativity, and distributivity and use them to compute with whole numbers.
- Express mathematical relationships using equations.
- Model problem situations with objects and use representations such as graphs, tables, and equations to draw conclusions.

Talk About It
Discuss the Try It! activity.

- **Say:** *Explain how the Commutative Property of Multiplication is modeled by the different numbers and colors of rods.*

- **Ask:** *How do the parentheses in $3 \times (6 \times 2)$ and $(3 \times 6) \times 2$ change the way you do the multiplication?* Talk about the Associative Property of Multiplication with students.

Solve It
Reread the problem with students. Have students explain which property is shown by each model.

More Ideas
For other ways to teach properties of multiplication—

- Use color tiles to model each of the three properties of multiplication. Have students discover the characteristics of each property and record a copy of the model and the related number sentences for each property on inch grid paper.

- Use Snap Cubes® to model various problems. Write number sentences on the board showing each property, such as $4 \times 5 = 5 \times 4$. Have students compare the areas (or lengths) in each model to see that they are equal. Explain which property is illustrated by each model.

Standardized Practice
Have students try the following problem.

Which number sentence shows the Commutative Property of Multiplication?

A. $18 \times 12 = 12 \times 18$ **C.** $8 \times (4 \times 5) = (8 \times 4) \times 5$

B. $6 \times 1 = 1$ **D.** $(11 + 7) + 25 = 11 + (7 + 25)$

Try It! 40 minutes | Groups of 4

Here is a problem about the properties of multiplication.

Sara and Ben have 3 dogs. Each dog gets 6 biscuit treats a week. Sara and Ben use different ways to calculate how many treats the dogs get altogether in a week. They also use different ways to calculate how many treats the dogs get in 2 weeks. Show the ways that Sara and Ben might use.

Introduce the problem. Then have students do the activity to solve the problem. Distribute the Cuisenaire® Rods, grid paper, paper, and colored pencils to students.

Materials
- Cuisenaire® Rods (2 sets per group)
- Centimeter Grid Paper (BLM 8; 1 per group)
- paper (1 sheet per group)
- colored pencils (2 per group)

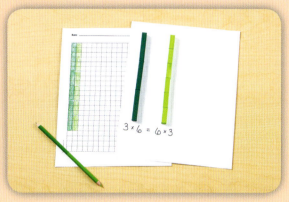

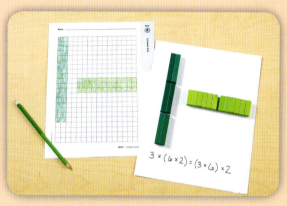

1. Have students model the two ways to find the number of treats all the dogs get in a week. Multiply the dogs first, and then multiply the treats first. Ask them to shade and then compare the two models on the grid paper. Write $3 \times 6 = 6 \times 3$ on the board. Explain that the models show the Commutative Property of Multiplication.

2. Write $3 \times (6 \times 2)$ and $(3 \times 6) \times 2$ on the board. **Say:** *Sara and Ben use different ways to find the number of treats the dogs get in 2 weeks. Model both ways.* Ask students to shade and then compare the models by counting the squares covered by rods on the grid paper. Introduce the Associative Property of Multiplication.

⚠ Look Out!

Some students may generalize the multiplication properties to include division. Have them model division using Cuisenaire® Rods to discover that division has different results when the order and/or grouping of the numbers is changed. Other students may try to use the Associative Property of Multiplication to group different operations. Stress that only one operation pertains to the property, as stated in the full name.

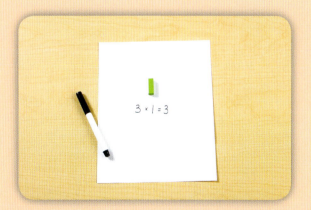

3. Say: *If one dog gets 3 treats a week, how many treats would Sara and Ben give it in one week?* Write 3×1 on the board. Have students model the expression. **Ask:** *What is the product of one and any number?* Define the Identity Property of Multiplication.

97

Lesson 3

Algebra

Distributive Property

Students who understand the Distributive Property have a powerful tool for computing mentally. Fluency with place value enables students to more easily decompose numbers. Knowledge of the use of parentheses helps students understand and compare symbolic representations. The Distributive Property states $a \times (b + c) = (a \times b) + (a \times c)$ and $a \times (b - c) = (a \times b) - (a \times c)$.

Try It! Perform the Try It! activity on the next page.

Objective

Identify and use the Distributive Property.

Skills

- Adding
- Multiplying
- Representing numbers

NCTM Expectations

Grades 3–5
Algebra
- Identify such properties as commutativity, associativity, and distributivity and use them to compute with whole numbers.
- Express mathematical relationships using equations.
- Model problem situations with objects and use representations such as graphs, tables, and equations to draw conclusions.

Talk About It

Discuss the Try It! activity.

- Discuss the common factor in the expression $(3 \times 10) + (3 \times 6)$.
- **Ask:** *How can you use the Distributive Property to help you find 3×16 mentally?*
- Have students describe how they would use the Distributive Property to rewrite 19×25. Have students identify which factor they would break apart and the common factor used in the expressions.

Solve It

Reread the problem with students. Have students explain in words, draw a model, and use numbers and symbols to show two possible ways to use the Distributive Property to find the solution.

More Ideas

For other ways to teach about the Distributive Property—

- Have students use base ten blocks to model the Distributive Property for products of larger numbers. For example, 3×197.
- Write 4×13 on the board. Have students use two-color counters to make a 4×13 array. Have them turn over counters in the last three columns to make a 4×3 array within the 4×13 array. Have students write the expressions shown by each array. Explain how the model shows the Distributive Property.

Standardized Practice

Have students try the following problem.

Which is one way to solve 7×28 using mental math?

A. $(7 \times 20) - (7 \times 2)$ **C.** $(7 \times 30) + (7 \times 2)$

B. $(7 \times 20) + (7 \times 8)$ **D.** $(7 + 20) \times (7 + 8)$

Try It! 20 minutes | Pairs

Here is a problem about the Distributive Property.

Micah buys 3 containers of seedlings. Each container has 16 seedlings. How many seedlings does Micah buy?

Introduce the problem. Then have students do the activity to solve the problem. Discuss what operation and expression could be used to solve the problem. Remind students of the meaning of parentheses in an expression and how they are used when computing. Distribute base ten blocks, paper, and pencils to students.

Materials
- base ten blocks (50 units and 10 rods per pair)
- paper (1 sheet per pair)
- pencils (1 per pair)

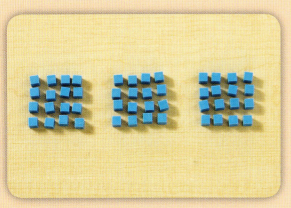

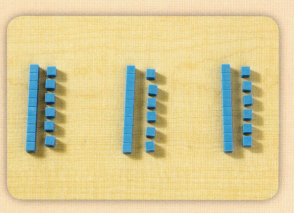

1. Say: *Micah buys 3 containers of 16 seedlings.* Have students model three groups of 16 using units. Write 3 × 16 on the board.

2. Ask: *What is another way to express 16?* Have students exchange 10 units for 1 rod in each group to show *10 + 6 = 16* three times. Write *3 × (10 + 6)* on the board.

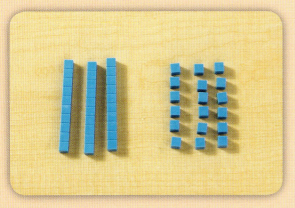

3. Ask: *How can you determine how many seedlings there are in all?* Have students separate the rods from the units and write an equation to represent the new model. Write *(3 × 10) + (3 × 6) = 30 + 18 = 48* on the board. Introduce the Distributive Property.

⚠ Look Out!

Watch for students who mistakenly use the incorrect sign when writing number sentences. For example, they may write *3 × 16 = (3 + 10) × (3 + 6)*. Suggest that these students model the problem and write down each step as they show it with blocks. Also, watch for students who don't distribute the factor all the way through the parentheses, for example, multiplying 3 × 10 but not 3 × 6 in the equation: *3 × (10 + 6)*.

Lesson 4

Algebra
Order of Operations

The order of operations makes the language of mathematics more universal. Knowing these rules helps students to communicate more accurately as they gain fluency in manipulating symbolic relationships. The sequence for the order of operations is listed below.

1. Calculate inside parentheses.
2. Multiply and divide in order, from left to right.
3. Add and subtract in order, from left to right.

Try It! *Perform the Try It! activity on the next page.*

Objective
Use the order of operations to simplify expressions.

Skills
- Adding and subtracting
- Multiplying and dividing
- Using parentheses to group numbers

NCTM Expectations

Grades 3–5
Algebra
- Model problem situations with objects and use representations such as graphs, tables, and equations to draw conclusions.

Number and Operations
- Understand and use properties of operations, such as the distributivity of multiplication over addition.

Talk About It
Discuss the Try It! activity.
- **Ask:** *Why do the models have different solutions?*
- **Ask:** *Why is it necessary to follow the order of operations when simplifying an expression?*
- Write $5 + 2 \times 6 - 8$ on the board. **Ask:** *How does the value of this expression differ when using the order of operations versus solving from left to right? Explain.*

Solve It
Reread the problem with students. Have students draw a picture of the solution to the problem. Then have them write a short paragraph explaining how to use the order of operations to solve the problem.

More Ideas
For other ways to teach the order of operations—
- Write $20 - 12 \div 4$ on the board. Have students use Snap Cubes® to model the expression and compute using the order of operations. Repeat with other expressions.
- Use two-color counters to model the problem: $5 - 3 + 6 \div 2 = 4$. Have students use the counters to help them decide where parentheses should be inserted in the equation.

Standardized Practice
Have students try the following problem.

Simplify $20 - 8 \div 4 \times 2$.

A. 1.5   B. 6   C. 9   D. 16

Try It! 20 minutes | Groups of 4

Here is a problem about the order of operations.

Jay brought some juice boxes to soccer practice to share with his teammates. He had 3 single boxes and 4 multi-packs. There are 6 single boxes in each multi-pack. To determine how many boxes of juice Jay brought to practice, evaluate $3 + 4 \times 6$.

Introduce the problem. Then have students do the activity to solve the problem. Distribute color tiles, paper, and pencils to students. Explain that the order of operations provides rules for simplifying expressions. Discuss the rules.

Materials
- color tiles (100 per group)
- paper (1 sheet per group)
- pencils (1 per group)

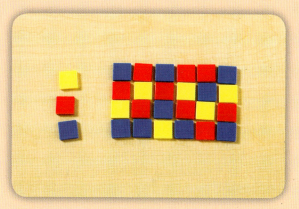

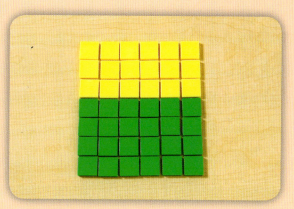

1. Write $3 + 4 \times 6$ on the board. Have students start by laying down 3 tiles. Then have students add a 4-by-6 array. **Ask:** *How many tiles are shown in the model?*

2. Have students show $3 + 4$ using a different color of tile for each addend. Then have them build an array to show this quantity times six. **Ask:** *How many tiles are shown in the model?*

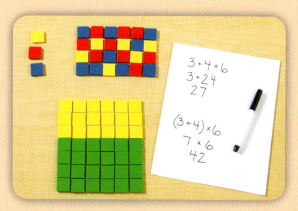

3. Say: *You built two models.* **Ask:** *How are they different?* Have students write the expressions to represent the models. **Ask:** *Which model is correct?*

⚠ Look Out!

Because we read English from left to right, some students may continue to simplify expressions by performing operations in that order. Suggest that students write the order of operations at the top of their papers and then refer to the steps as they simplify expressions. Some students find a mnemonic, such as *Please My Dear Aunt Sally (Parentheses Multiplication Division Addition Subtraction)*, helpful in remembering the order.

Lesson 5

Algebra
Expressions with a Variable

As students become more fluent in computation, they begin to understand that mathematical relationships are not always static. For example, most students recognize that the total cost of movie tickets depends upon the number of tickets purchased. Using a variable as a placeholder for an unknown value allows them to communicate this relationship between cost and quantity. Students should recognize that variables are any letter or symbol used to represent a number.

Try It! Perform the Try It! activity on the next page.

Objective
Write and evaluate an expression with a variable.

Skills
- Multiplying
- Representing numbers

NCTM Expectations

Grades 6–8
Algebra
- Develop an initial conceptual understanding of different uses of variables.
- Use symbolic algebra to represent situations and to solve problems, especially those that involve linear relationships.
- Model and solve contextualized problems using various representations, such as graphs, tables, and equations.

Talk About It
Discuss the Try It! activity.

- **Ask:** *How can the expressions $n \times 5$ and $c \times 5$ describe the same situation?*
- **Say:** *I have some green tiles and six red tiles in my hand.* **Ask:** *What expression describes how many tiles I have?* Have students suggest other situations that can be described using variable expressions. Elicit examples for addition, subtraction, multiplication, and division.
- Have students explain how to evaluate $x - 3$ for $x = 10, 28,$ and 52.

Solve It
Reread the problem with students. Have them write a paragraph explaining the meaning of the expression and how they can use models and symbols to evaluate the expression for the sixth car.

More Ideas
For other ways to teach about expressions with variables —

- Write $a - 2$, $b + 2$, $2c$, and $d \div 2$ on the board. Have students give different ways to read each expression, such as *a minus 2* and *2 less than a* for $a - 2$. Have students use two-color counters to model the four expressions when the variable in each equals 4 and when it equals 10. Discuss how to substitute numbers to evaluate expressions symbolically.
- Use polyhedra dice to provide students with more practice. Have students write an expression, state different ways to read the expression, and then roll the dice. They use the rolled number to evaluate the expression.

Standardized Practice
Have students try the following problem.

Evaluate $33 + z$ for $z = 11$.

A. 3   B. 11   C. 22   D. 44

Try It! 30 minutes | Groups of 4

Here is a problem about expressions with a variable.

It takes 5 minutes to wash each car at Details Car Wash. The total time needed to get a car washed depends upon the car's position in line. Write an expression to show the number of minutes it will take Sally to get her car washed. Then evaluate the expression to determine how long it takes Sally to get her car washed if her car is the sixth car in line.

Introduce the problem. Then have students do the activity to solve the problem. Distribute color tiles, paper, and pencils to students. Write 1×2 and $n \times 2$ on the board. Point out that each statement is an expression because it contains only numbers and/or symbols and operations (no equal sign).

Materials
- color tiles (100 per group)
- paper (3 sheets per group)
- pencils (1 per group)

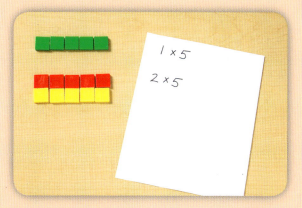

1. Say: *Let each tile represent one minute. Make an array to show how many minutes it will take Sally to get her car washed if she is the first car in line.* Have students write the expression shown by the array. Repeat for 2, 3, and 4 cars.

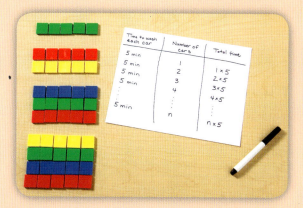

2. Have students create a table to organize the information shown by each array. **Say:** *What changes in each expression?* Write $n \times 5$ on the board. Introduce the term *variable*.

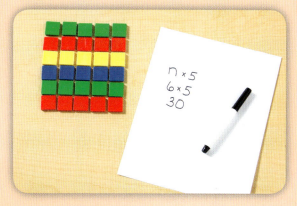

3. Say: *Find the length of the wait when n = 6.* Explain that finding the value of an expression is called *evaluating* the expression. Have students model the situation.

⚠ Look Out!

Students who have difficulty writing variable expressions may find it helpful to first think of the situation as static and use numbers to help them identify the relationship. Once they understand the relationship, they can replace the "made-up" number in the expression with a letter. Remind students that when they evaluate an expression, the operations and numbers (constants) in the expression do not change. Suggest that they think of the variable as the only amount that can vary, or change.

Lesson 6

Algebra
Equations with a Variable

Students further develop their mathematical flexibility by using variable expressions to write equations. They use concrete models and mental math to solve for the unknown value in an equation. In the activity, the implicit use of inverse operations involving only basic facts lays the foundation for using symbolic representations to solve more complex equations.

Objective
Write and solve an equation with a variable.

Skills
- Using operations
- Interpreting data
- Representing numbers

NCTM Expectations

Grades 3–5
Algebra
- Represent the idea of a variable as an unknown quantity using a letter or a symbol.
- Express mathematical relationships using equations.
- Model problem situations with objects and use representations such as graphs, tables, and equations to draw conclusions.

Number and Operations
- Develop fluency in adding, subtracting, multiplying, and dividing whole numbers.

Try It! *Perform the Try It! activity on the next page.*

Talk About It
Discuss the Try It! activity.
- **Ask:** *How does mental math help you to solve the equations?*
- **Ask:** *How do you solve $5 \times b = 20$ using a related fact?*
- **Say:** *Tell a story that could be solved using the equation $36 \div p = 4$. Explain how to solve the equation.*

Solve It
Reread the problem with students. Have them write a paragraph to explain how they use the models and mental math to solve the equations.

More Ideas
For other ways to teach about equations with a variable—
- Have students use two-color counters to model the problem. They can start with 20 counters, all with the red side showing, to represent the apples that Bryce had after he went to the store. Students then turn over 6 of the counters to represent the apples that he bought at the store. Guide students to use the remaining red counters to solve the variable equation.
- Have groups of students make up simple word problems, use centimeter cubes to model the data, and write equations that can be used to solve the problem. Encourage students to use mental math to solve each problem.

Standardized Practice
Have students try the following problem.

There are 24 coins on Miguel's desk. Of these, 6 coins are quarters and the rest are dimes. Which equation could be used to find the number of dimes on Miguel's desk?

A. $6 \times d = 24$ **B.** $d - 24 = 6$ **C.** $d + 6 = 24$ **D.** $d \div 6 = 24$

Try It! 20 minutes | Groups of 4

Here is a problem about equations with a variable.

Bryce has some apples. He buys 6 more at the store. Now he has 20 apples. Write an equation to determine how many apples Bryce had before he went to the store. He has a recipe that uses 5 apples to make one batch of applesauce. Write an equation to determine how many batches of applesauce he can make with 20 apples.

Introduce the problem. Then have students do the activity to solve the problem. Distribute the Cuisenaire® Rods, paper, and pencils to students. Explain that an equation is a statement that two quantities are equal.

Materials
- Cuisenaire® Rods (1 set per group)
- paper (2 sheets per group)
- pencils (1 per group)

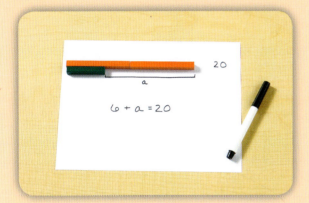

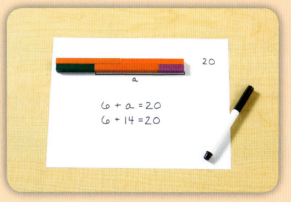

1. Have students state what is unknown. Using rods, have students model the facts of the first question. **Ask:** *What operation does buying more apples suggest?* Guide students to write an addition equation using a variable to represent the unknown number of apples. Write $6 + a = 20$ on the board.

2. Have students complete the model by making the two rows equal. **Say:** *Think "Six plus what number equals 20?"* Encourage students to use mental math or related facts when they solve the equation.

⚠ Look Out!

Students who cannot write equations may need to list the known facts and the information needed to solve the problem. Review key words that may help students recognize the operation needed to solve the problem. Reviewing and modeling fact families to see the relationships between operations may benefit the students.

3. Say: *Bryce uses 5 apples to make one batch of applesauce. What equation would you use to determine how many batches of applesauce he can make with 20 apples?* Have students model the problem and write the multiplication equation.

Lesson 7

Algebra
Addition and Subtraction Equations

Students who first solve simple, intuitive equations, such as $x + 1 = 4$, and analyze how they find the solution, tend to understand the concepts for example, inverse operations, used in solving equations. Applying these methods to slightly more difficult equations helps students to expand and generalize their understanding.

Try It! Perform the Try It! activity on the next page.

Objective
Solve addition and subtraction equations.

Skills
- Adding and subtracting
- Using inverse relationships
- Representing numbers

NCTM Expectations

Grades 3–5
Algebra
- Express mathematical relationships using equations.
- Model problem situations with objects and use representations such as graphs, tables, and equations to draw conclusions.

Number and Operations
- Identify and use relationships between operations, such as division as the inverse of multiplication, to solve problems.
- Develop fluency in adding, subtracting, multiplying, and dividing whole numbers.

Talk About It
Discuss the Try It! activity.
- **Ask:** *How does subtraction help you solve the problem?*
- **Say:** *Choose a number. Add 6 to the number; then subtract 6 from the sum.* **Ask:** *What number is left?* Explain that *adding 6 and subtracting 6* are inverse operations since one action undoes the other. Have students give other examples.

Solve It
Reread the problem with students. Have them explain how to use inverse operations to solve the equation. Then have students write the solution to the addition equation in the proper form.

More Ideas
For other ways to teach about solving addition and subtraction equations—
- Use two-color counters to model the problem. Have students fold a sheet of paper in half with each half representing one side of the equation. Have them add or subtract counters on one side to make the two sides equal. Discuss how to use inverse relationships to find the solution.
- Have each student make up two addition and two subtraction equations that each contain one variable. Students then use Snap Cubes® to model each equation. Have them write one related subtraction equation for each addition equation and one related addition equation for each subtraction equation.

Standardized Practice
Have students try the following problem.

Which equation expresses the solution to $k - 8 = 16$?

A. $16 - 8 = 8$ **C.** $16 \div 2 = 8$

B. $16 + 8 = 24$ **D.** $2 \times 8 = 16$

Try It! 20 minutes | Groups of 4

Here is a problem about solving addition and subtraction equations.

Randy spent 39 minutes on his computer. He spent the first 22 minutes downloading music and the remaining time messaging friends. Write and solve an addition equation to determine how many minutes Randy spent messaging his friends.

Introduce the problem. Then have students do the activity to solve the problem. Distribute Cuisenaire® Rods, paper, and pencils to students. Suggest that m represents the number of minutes that Randy spent messaging his friends.

Materials
- Cuisenaire® Rods (2 sets per group)
- paper (1 sheet per group)
- pencils (1 per group)

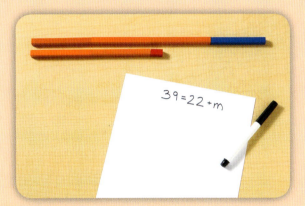

1. Write $39 = 22 + m$ on the board **Ask:** *How does this equation represent the situation?* Have students use Cuisenaire Rods to model the equation.

2. Say: *To solve this equation, you can think "39 equals 22 plus what number?"* Have students complete the model and write the equation that represents the model.

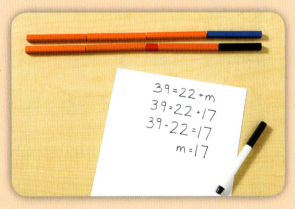

3. Have students refer to the model to visualize the inverse relationship between addition and subtraction. Elicit that students can subtract 22 from 39 to solve for m. **Say:** *Write a subtraction equation that expresses the information in the problem.*

Look Out!

Be sure students left-align the rods in their models to help them recognize the relationship between the numbers.

Lesson 8

Algebra
Multiplication and Division Equations

Students use variables as placeholders for missing numbers in equations. Arrays demonstrate the inverse relationship between multiplication and division and provide a foundation for the use of symbolic representations to solve equations. To be fluent, students must recognize that a relationship may be expressed in different ways.

Try It! Perform the Try It! activity on the next page.

Objective
Solve multiplication and division equations.

Skills
- Multiplying and dividing
- Using inverse relationships
- Representing numbers

NCTM Expectations

Grades 3–5
Algebra
- Express mathematical relationships using equations.
- Model problem situations with objects and use representations such as graphs, tables, and equations to draw conclusions.

Number and Operations
- Identify and use relationships between operations, such as division as the inverse of multiplication, to solve problems.
- Develop fluency in adding, subtracting, multiplying, and dividing whole numbers.

Talk About It
Discuss the Try It! activity.
- **Ask:** *Why can you rewrite a multiplication equation as a division equation?*
- **Ask:** *What does it mean when we say that division is the inverse of multiplication?*
- **Ask:** *How can you solve $p \div 7 = 8$ using multiplication?*

Solve It
Reread the problem with students. Ask them to explain in writing how they used the relationship between multiplication and division to solve the problem. Have students use the models to help them. Remind students to write the solution to the multiplication equation.

More Ideas
For other ways to teach solving multiplication and division equations—
- Use color tiles to make arrays to solve the problem. Some students may prefer to organize their work by counting out 32 tiles and placing each individual tile in the array.
- Use two-color counters to model related multiplication and division equations. Provide students with equations such as $6 \times a = 66$, $b \times 4 = 56$, $d \div 2 = 14$, and $48 \div e = 16$. Have students make stacks of counters to show the multiplication and division equations. Discuss how to use inverse relationships to find the solutions to the equations.

Standardized Practice
Have students try the following problem.

Which equation expresses the solution to $n \times 5 = 60$?

A. $60 - 5 = 55$
B. $5 + 60 = 65$
C. $5 \times 60 = 300$
D. $60 \div 5 = 12$

Try It! 20 minutes | Groups of 4

Here is a problem about solving multiplication and division equations.

Zoe has 4 bookshelves. If each shelf has the same number of books, and there are 32 books in all, how many books are on each shelf? Write a multiplication equation to represent the number of books on each shelf. Use division to solve the equation.

Introduce the problem. Then have students do the activity to solve the problem. Distribute Cuisenaire® Rods, paper, and pencils to students. Suggest that b represent the number of books on each shelf.

Materials
- Cuisenaire® Rods (2 sets per group)
- paper (1 sheet per group)
- pencils (1 per group)

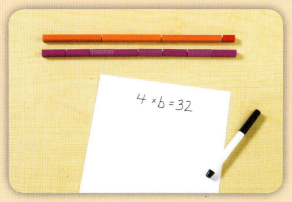

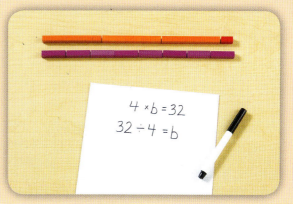

1. Write $4 \times b = 32$ on the board and have students discuss how the equation is related to the problem. Have students use rods to model the equation.

2. Ask: *How can you solve the problem using division? What division equation is shown by the model?* Write $32 \div 4 = b$ on the board.

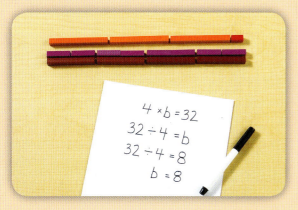

3. Have students solve for *b*. Have them discuss what each purple rod represents. Tell students to use brown rods to make an alternate version of the model, and ask them to discuss what each brown rod represents.

⚠ Look Out!

If students have trouble determining the division equation, they may find it helpful to review multiplication and division fact families. Have them write the four facts that use the numbers *3*, *4*, and *12*. Then have them compare the four facts to see how they are alike and how they are different. Have students make a 3×4 array and explain how each fact is shown by the array.

Lesson 9

Algebra
Patterns and Function Tables

The ability to recognize and use patterns to solve problems forms the basis of algebraic thinking. Representing patterns in function tables allows students to discover patterns more easily and to see how x-values relate to y-values. These relationships can then be used to make predictions. Function tables will help students to graph relationships between x and y.

Try It! Perform the Try It! activity on the next page.

Objective
Use patterns and function tables to solve problems.

Skills
- Representing patterns
- Identifying functions
- Understanding patterns and functions

NCTM Expectations
Grades 3–5
Algebra
- Describe, extend, and make generalizations about geometric numeric patterns.
- Represent and analyze patterns and functions, using words, tables, and graphs.
- Model problem situations with objects and use representations such as graphs, tables, and equations to draw conclusions.

Talk About It
Discuss the Try It! activity.

- **Ask:** *What does the variable x in the function table represent? What does the variable y represent?*
- **Ask:** *What words can you use to describe the pattern in the function table? How does the function table help you see the pattern?*
- **Ask:** *Why should you test the function rule for all of the values in the table?*

Solve It
Reread the problem with students. After they build a model for 4 stones, students can use the pattern in the table to predict the number of bricks for 6 stones. Have students complete the function table to check their predictions.

More Ideas
For other ways to teach about using patterns and function tables—

- Have students use triangle pattern blocks to find the perimeter of one triangle, then two, three, and four triangles placed side-to-side. They record the number of triangles as x in the function table and the perimeter as y. Have them predict perimeters for 6, 7, and 8 triangles placed side-to-side.
- Have students use centimeter cubes to build other patterns, such as triangular numbers. Guide them to use 1, 3, 6, and 10 centimeter cubes to build triangles. Have them use the pattern in the function table to predict the number of cubes used in the fifth figure.

Standardized Practice
Have students try the following problem.

Raul is using blue and gold tiles to create a mosaic pattern. If x represents blue tiles and y represents gold tiles, how many gold tiles does Raul use if he uses 8 blue tiles?

A. 11 C. 21
B. 18 D. 24

x	y
3	9
4	12
5	15

Try It! 20 minutes | Pairs

Here is a problem about patterns and function tables.

Lee Ann will use square and triangular stones to make a path in her garden. Each side of a triangular stone is the same length as each side of a square stone. The triangles will be used as a border around a row of squares. For example, she would use four triangles for a border around one square and six triangles for a border around two squares. How many triangles will she need for a border around six squares?

Introduce the problem. Then have students do the activity to solve the problem. Distribute pattern blocks, Function Tables, and pencils to students.

Materials
- pattern blocks (25 per pair)
- Function Tables (BLM 10; 1 per pair)
- pencils (1 per pair)

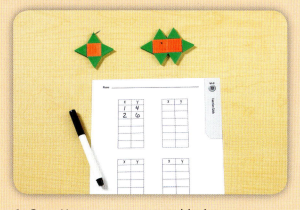

1. Say: *Use a square pattern block to represent each square stone. Use a triangle to represent each triangular stone.* Have students create models for paths with one and two squares. Then for each path have them record the number of squares in the *x* column and the number of triangles in the *y* column.

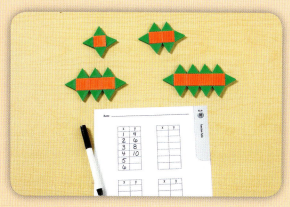

2. Say: *Use more squares to model paths with three and four square stones. Add triangles and then fill in the next two rows of the table.* Have students identify the pattern and use it to fill in the table for 5 and 6 square stones.

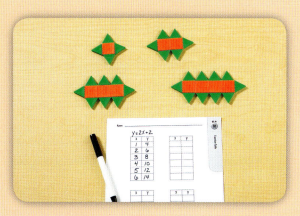

3. Ask: *What operations can you perform on the x-values in the table to get the y-values?* Guide students to see that they can multiply *x* by 2 and add 2 to get *y*. Have them write the equation as $y = 2x + 2$ and then check that the equation works for all of the values in the table.

⚠ Look Out!

Some students may have difficulty finding a rule to represent the pattern in the function table. Have them look at the model for two squares placed end-to-end. Point out that the number of triangles on each side of a square is twice the number of squares since there are two sides to each square, in addition there is a triangle at each end. Have them check that this pattern holds for each number of square stones.

Lesson 10

Algebra
Introduction to Integers

Students are typically introduced to negative numbers using temperatures below zero, losses in games, and money owed. The concepts of *magnitude, direction, and opposites* are important for understanding integers. Students can think about an integer as a distance (the magnitude, or absolute value) in either direction on the number line. Numbers to the right of zero are positive, and numbers to the left of zero are negative. A number some distance from zero in one direction is the opposite of the number the same distance from zero in the other direction.

Try It! *Perform the Try It! activity on the next page.*

Objective
Represent, compare, and order integers.

Skills
- Representing integers
- Comparing and ordering integers
- Reasoning

NCTM Expectations
Grades 6–8
Algebra
- Relate and compare different forms of representation for a relationship.

Number and Operations
- Develop meaning for integers and represent and compare quantities with them.

Talk About It
Discuss the Try It! activity.

- *Discuss the correct way to say negative numbers—for example, negative four for -4.* **Ask:** *What situations can you represent using negative integers?*

- **Ask:** *What is the opposite of 3? –4? 0?* Have students represent the integers and their opposites on a number line. Point out that 0 is neither positive nor negative and is its own opposite.

- **Ask:** *Is it possible for a positive integer to be less than a negative integer? Why or why not?* **Say:** *Explain why –4 is less than –1.*

Solve It
Reread the problem with students. Have them explain how they know that 3 is the first-place score and –4 is the last-place score for the round.

More Ideas
For other ways to teach integers—

- Have students use a vertical number line (BLM 11 in portrait orientation) to show temperature or altitude, with tick marks representing increments of 1, 2, 5, or 10. With the number line as a guide, students use color tiles to represent temperatures or altitudes greater than or less than a given value. Have students use < or > to demonstrate comparisons between the integers.

- Have students use an Integer Number Line (BLM 11) with only certain numbers labeled, such as –4, 0, and 4. Have them use two-color counters to locate various integers on the number line, such as –3 or the opposite of 2. Have students write the integers below the tick marks.

Standardized Practice
Have students try the following problem.

Which of the following is the least altitude?

A. –32 meters   B. –8 meters   C. –3 meters   D. 2 meters

Try It! 20 Minutes | Pairs

Here is a problem about representing and ordering integers.

Dustin, Kyle, and Emma are playing a word game. After one round of play, Dustin lost 4 points, Emma gained 3 points, and Kyle lost 1 point. How can you use integers to represent the scores for this round of play? Which scores represent first place and last place?

Introduce the problem. Then have students do the activity to solve the problem. Distribute color tiles, number lines sheets, paper, and pencils to students. Draw a number line from −4 to 4 on the board. Label the tick marks. Have students copy the numbering onto the number lines on the sheet.

Materials
- color tiles (11 per pair)
- 1-Inch Number Lines (BLM 11; 1 per pair)
- paper (1 sheet per pair)
- pencils (1 per pair)

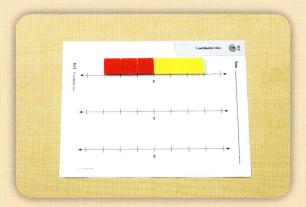

1. Say: *Integers are the counting numbers, their opposites, and 0.* Have students place 3 red tiles to the left of 0 and 3 yellow tiles to the right of 0 on the number line. **Say:** *−3 and 3 are opposites because they are the same distance from 0 on the number line.*

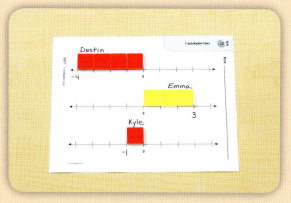

2. Say: *Negative integers can represent losses, below-zero temperatures, and below-sea-level altitudes.* **Ask:** *What do positive integers represent?* Have students represent the losses and gains from the game on the number lines.

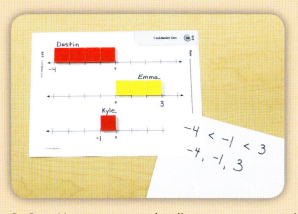

3. Say: *You can use number lines to compare and order integers. The values of the numbers increase as you move left to right.* Have students use symbols to compare the three integers from the game. Then have them write the integers in order from least to greatest.

⚠ Look Out!

Some students may be confused because an integer such as −8 is less than −1. Use real-world contexts to help them see that the farther a negative integer is from 0, the less is its value. For example, the person who owes $8 has less money than the person who owes $1, or −8°F is below −1°F on the thermometer.

Lesson 11

Algebra
Add Integers

Addition is the same for whole numbers and integers—the grouping together of quantities. Modeling with different colors helps students to perform operations with positive and negative numbers. As students learn the rules for working with integers they should make the connection between the models they build and the rules so that manipulating integers is not arbitrary.

Objective
Add integers.

Skills
- Adding integers
- Representing integers
- Understanding addition

NCTM Expectations

Grades 6–8
Algebra
- Relate and compare different forms of representation for a relationship.

Number and Operations
- Understand the meaning and effects of arithmetic operations with fractions, decimals, and integers.
- Develop and analyze algorithms for computing with fractions, decimals, and integers and develop fluency in their use.

Try It! Perform the Try It! activity on the next page.

Talk About It
Discuss the Try It! activity.

- **Ask:** *After you model the problem with two-color counters, why do you rearrange them to form red-yellow pairs? What does a red-yellow pair represent in terms of yardage?* Students should recognize that one yard lost (red) plus one yard gained (yellow) is a net change of zero, so a red-yellow pair represents no gain or loss.

- **Ask:** *Is the sum of two negative numbers always negative? Model an example to justify your answer.*

Solve It
Reread the problem with students. Since 12 > 5, the team gained more yards than it lost. Since 12 − 5 = 7, they gained 7 more yards than they lost. So their net yardage is 7 yards. Have students write −5 + 12 = 7 and explain how this equation relates to the problem.

More Ideas
For other ways to teach about adding integers—

- Have students use centimeter cubes and a 1-cm Number Line (BLM 11) to add pairs of integers—two positive numbers: 2 + 6, two negative numbers: −1 + (−5), and a positive number and negative number: −9 + 2 or 8 + (−4). Suggest that students use red cubes for negative numbers and yellow cubes for positive numbers.

- Ask students to write and model their own number sentences using color tiles to show that a positive number plus a negative number can be positive, negative, or zero.

Standardized Practice
Have students try the following problem.

The morning temperature of −9°F is expected to rise 10 degrees by noon. What is the expected noon temperature?

A. −19°F   B. −1°F   C. 1°F   D. 19°F

Try It! 15 minutes | Pairs

Here is a problem about adding integers.

A football team lost 5 yards on one play, then gained 12 yards on the next play. What was the team's net yardage on the two plays?

Introduce the problem. Then have students do the activity to solve the problem. Distribute two-color counters, number lines sheets, and pencils to students.

Materials
- two-color counters (at least 20 per pair)
- 1-cm Number Lines (BLM 12; 1 per pair)
- pencils (1 per pair)

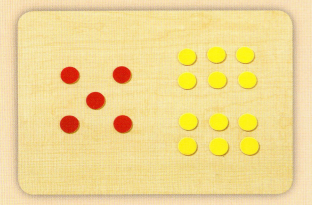

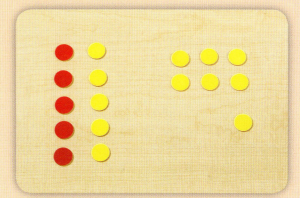

1. Say: *Each red counter represents one yard lost. Each yellow counter represents one yard gained. Place counters to model this problem.* Students place 5 red counters and 12 yellow counters.

2. Say: *Move counters so that each red counter is paired with a yellow counter. What number does each red-yellow pair represent? How many yellow counters are left?* Students form 5 pairs, each representing 0. There are 7 yellow counters left, representing a net gain of 7 yards.

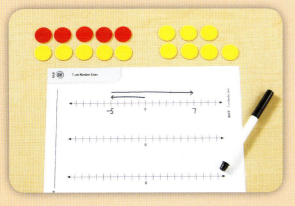

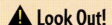

⚠ Look Out!

Students may confuse negative signs with minus signs. Have students write the problem $-5 + 12 = 7$. Then speak the correct words: negative five (not minus five) *plus twelve equals seven*. On a number line, show students that -5 and 5 are opposites. They are both 5 units from 0, but in opposite directions.

3. Say: *Now use a number line to solve this problem. Starting at 0, draw a segment 5 units to the left. From −5, draw a segment 12 units to the right.* **Ask:** *At what number do you end?* Students draw the two segments and end at 7. Help students recognize that the overlapping parts of the lines are equivalent to red-yellow pairs of counters.

Lesson 12

Algebra
Subtract Integers

After students show confidence with adding integers, they can learn to subtract integers. They will continue to use and develop their understanding of addition and subtraction as inverse operations. Previous work with fact families will help students to think flexibly as they add and subtract positive and negative numbers.

Objective
Subtract integers.

Skills
- Subtracting integers
- Representing integers
- Using inverse operations

NCTM Expectations

Grades 6–8
Algebra
- Relate and compare different forms of representation for a relationship.

Number and Operations
- Understand and use the inverse relationships of addition and subtraction, multiplication and division, and squaring and finding square roots to simplify computations and solve problems.
- Understand the meaning and effects of arithmetic operations with fractions, decimals, and integers.
- Develop and analyze algorithms for computing with fractions, decimals, and integers and develop fluency in their use.

Try It! *Perform the Try It! activity on the next page.*

Talk About It
Discuss the Try It! activity.

- **Ask:** *Is this problem about a subtraction problem or an addition problem?* Discuss with students.
- **Ask:** *When you think about this problem in two different ways—as addition and as subtraction—do you get two different answers?*
- Have students write the two number sentences for this problem.

Solve It
Reread the problem with students. Notice that only red counters are used to solve the subtraction problem, –8 – (–6) = –2. Both yellow and red counters are used to solve the addition problem, –8 + 6 = –2. Make sure students understand both ways to think about this problem. Either way, Hannah still owes Rachel $2 at the end.

More Ideas
For other ways to teach about subtracting integers—

- Students can use red and yellow color tiles to model the problem.
- Have students use centimeter cubes to find –2 – (–5). Suggest that students use red cubes for negative numbers and yellow cubes for positive numbers. They start with 2 red cubes and need to take away 5. But there are only 2 cubes available to take away, so 3 red-yellow pairs (which equal 0) can be added. Then 5 red cubes are removed, and 3 yellow cubes are left.

Standardized Practice
Have students try the following problem.

The current temperature is –6°F and is expected to drop 10 degrees overnight. What is the expected low temperature overnight?

A. –16°F   B. –10°F   C. –4°F   D. 4°F

Try It! 15 Minutes | Pairs

Here is a problem about subtracting integers.

At the bookstore, Hannah borrowed $8 from her sister Rachel. At the waterpark a few days later, Rachel borrowed $6 from Hannah. What is Hannah's standing with Rachel now? Does Hannah still owe Rachel any money?

Introduce the problem. Then have students do the activity to solve the problem. Distribute two-color counters to students.

Materials
- two-color counters (at least 20 per pair)

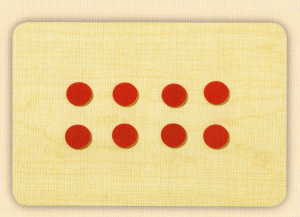

1. Say: *Let each red counter represent one dollar owed, or negative one. Use counters to show Hannah's situation after borrowing $8 from Rachel.* Students place 8 red counters on the table.

2. Say: *Later, Rachel borrowed $6 from Hannah. One way to think of this is that $6 of Hannah's debt to Rachel is taken away. This is a subtraction problem: –8 – (–6). Show this with the counters.* Students take away 6 red counters, and 2 are left.

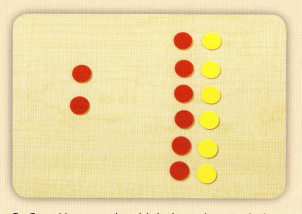

3. Say: *You can also think that when Rachel borrowed $6 from Hannah, it was the same as Hannah paying $6 back to Rachel. It is an addition problem: –8 + 6. Show this with the counters.* Students place 8 red counters, then add 6 yellow counters. They form 6 red-yellow pairs, and 2 red counters are left.

⚠ Look Out!

Students often get confused when they try to subtract a negative number, as in –8 – (–6). When they *take away* 6 red counters from a set of 8, students see that they can actually subtract a negative number. In this activity they also see that subtracting negative 6 is the same as adding positive 6: –8 – (–6) = –8 + 6. Once students are convinced, encourage them to use this concept anytime they see two negative signs together. For example, 1 – (–4) = 1 + 4 = 5.

Algebra
Multiply Integers

Students have developed the meaning of multiplication of whole numbers by using representations such as equal-sized groups, arrays, area models, and equal jumps on a number line. Some of these representations also work for multiplication with negative numbers. Understanding multiplication of integers prepares students for division of integers.

Try It! Perform the Try It! activity on the next page.

Objective
Multiply integers.

Skills
- Multiplying integers
- Representing integers
- Understanding multiplication

NCTM Expectations

Grades 6–8
Algebra
- Relate and compare different forms of representation for a relationship.

Number and Operations
- Understand the meaning and effects of arithmetic operations with fractions, decimals, and integers.
- Develop and analyze algorithms for computing with fractions, decimals, and integers and develop fluency in their use.

Talk About It
Discuss the Try It! activity.

- **Say:** When Ryan takes $5 out of his savings account, the integer –5 is used to describe the change in the amount of money in the account. **Ask:** When Ryan donates $5 to the food bank, what integer describes the change in the amount of money the food bank has?

- **Say:** The multiplication in this problem is $3 \times (-5)$. Compare this with 3×5. **Ask:** How are they the same? How are they different?

Solve It
Reread the problem with students. The amount of money in Ryan's savings account decreases each Friday, so a negative number (–5) is used to represent the change. To show the change in Ryan's account after 3 Fridays, students model the equation $3 \times (-5) = -15$, Have students explain the model.

More Ideas
For other ways to teach about multiplying integers—

- Have students use yellow and red centimeter cubes to model this and similar problems.
- Summarize the rules for multiplying integers.

  (1) The product of two positive integers is positive.
  (2) The product of two negative integers is positive.
  (3) The product of a positive integer and a negative integer is negative.

  Using two-color counters, have students model each rule. To model the product of two negative numbers, guide students to use repeated subtraction. To subtract groups of negative quantities from zero, first add red-yellow pairs. Then take away the red counters as appropriate.

Standardized Practice
Have students try the following problem.

In a computer game, you can win a maximum of 50 points and lose a maximum of 25 points in each round. What is the lowest possible score after three rounds?

A. –150   B. –75   C. –50   D. –25

Try It! 15 minutes | Pairs

Here is a problem about multiplying integers.

Ryan has a savings account. Every Friday he takes out $5 from the account and donates the money to the local food bank. What is the change in Ryan's account after three Fridays?

Introduce the problem. Then have students do the activity to solve the problem. Distribute two-color counters, number lines sheets, and pencils to students.

Materials
- two-color counters (at least 20 per pair)
- $\frac{1}{2}$-cm Number Lines (BLM 13; 1 per pair)
- pencils (1 per pair)

1. Say: *Let each red counter represent one dollar donated—which is one dollar less in Ryan's savings account, or negative one. Use counters to show the change in Ryan's account when he makes one donation.* Students display 5 red counters.

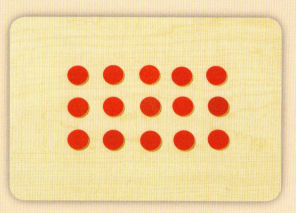

2. Say: *Now use counters to represent the change in Ryan's account after three Fridays. Organize the counters to show that there are three equal-sized groups.* Students display 3 groups of red counters, with 5 in each group. **Ask:** *What amount of money is represented?*

⚠ Look Out!

Sometimes students will be reluctant to think of multiplication as repeated addition when negative numbers are involved. Remind them that 3 × 5 is *3 groups of 5*, or 5 + 5 + 5, and that this idea applies to negative numbers, too. That is, 3 × (–5) is *three times negative five*, or *3 groups of –5*, or (–5) + (–5) + (–5).

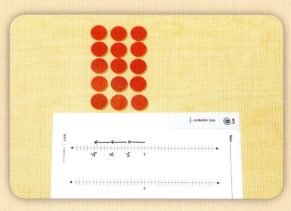

3. Say: *Now model this problem on a number line.* **Ask:** *How can you show that Ryan has $5 less in his savings account each Friday, for three Fridays?* Starting at 0, students jump 5 units left three times, ending at –15.

Lesson 14

Algebra
Divide Integers

Students can use what they already know about multiplying integers to divide integers. Multiplication and division are inverse operations. So, for example, to find the quotient 30 ÷ 6, students can think of the related product: 6 × ? = 30. The rules for division of positive and negative numbers are the same as those for multiplication.

Objective
Divide integers.

Skills
- Dividing integers
- Representing integers
- Using inverse operations

NCTM Expectations

Grades 6–8
Algebra
- Relate and compare different forms of representation for a relationship.

Number and Operations
- Understand the meaning and effects of arithmetic operations with fractions, decimals, and integers.
- Understand and use the inverse relationships of addition and subtraction, multiplication and division, and squaring and finding square roots to simplify computations and solve problems.
- Develop and analyze algorithms for computing with fractions, decimals, and integers and develop fluency in their use.

Try It! Perform the Try It! activity on the next page.

Talk About It
Discuss the Try It! activity.

- **Say:** *The addition problem for finding the sum of the scores is –4 + 1 + (–3).* **Ask:** *Does it matter which two numbers you add first?*
- **Say:** *The division problem for finding the average score is –6 ÷ 3 = –2. Write a related multiplication problem.* Students can write either 3 × (–2) = –6, or (–2) × 3 = –6.
- **Ask:** *Looking at the three scores, is it reasonable that the answer is negative rather than positive?* Have students explain their answers.

Solve It
Reread the problem with students. The average score is the sum of the scores divided by the number of scores: [–4 + 1 + (–3)] ÷ 3. The sum of the scores is –6 and –6 ÷ 3 is –2. Ken's average score is –2.

More Ideas
For other ways to teach about dividing integers—

- Have students use centimeter cubes to model pairs of number sentences, such as –10 ÷ 5 = –2 and 5 × (–2) = –10. Suggest that students use red cubes for negative numbers and yellow cubes for positive numbers. Also, discuss why –8 ÷ 4 is easier to model than 8 ÷ (–4).

- Summarize the rules for dividing integers, and note that they are the same as the rules for multiplying integers.

    (1) The quotient of two positive integers is positive.
    (2) The quotient of two negative integers is positive.
    (3) The quotient of a positive integer and a negative integer is negative.

    Using two-color counters, have students model an equation for rules 1 and 3.

Standardized Practice
Have students try the following problem.

Find the average of -6, -4, 8, and -2.

A. –4   B. –1   C. 1   D. 4

Try It! 15 minutes | Pairs

Here is a problem about dividing integers.

In three rounds of golf, Ken shot scores of −4, +1, and −3. What was his average score?

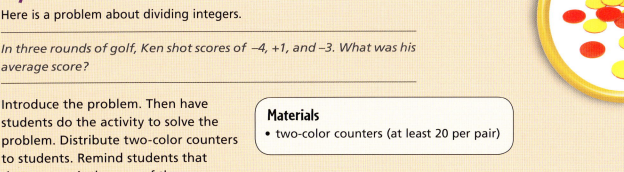

Introduce the problem. Then have students do the activity to solve the problem. Distribute two-color counters to students. Remind students that the average is the sum of the scores divided by the number of scores.

Materials
- two-color counters (at least 20 per pair)

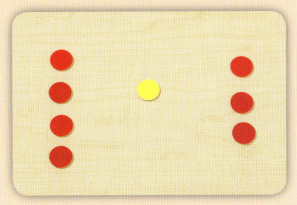

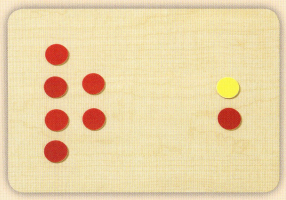

1. Say: *Each red counter represents negative one, and each yellow counter represents positive one. Use counters to show Ken's three scores.* Students place 4 red counters together, 1 yellow counter by itself, and 3 red counters together.

2. Say: *To find the average score, you first need to add the three scores. Use the counters to find the sum.* Students pair one yellow counter with a red counter to equal zero, and move the pair aside. There are 6 red counters left. The sum is −6.

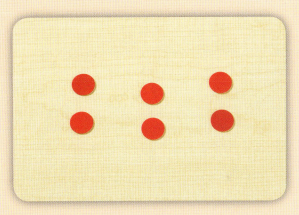

3. Say: *Now divide the sum by 3, since there are 3 scores. Divide the counters that represent the sum into 3 equal groups.* **Ask:** *How many counters are in each group?* Students arrange the 6 red counters into 3 groups, with 2 red counters in each group.

⚠ Look Out!

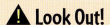

Make sure students understand that the average does not have to be one of the scores. To calculate the average score, they must add all the scores and divide by the number of scores, even though the scores include both positive and negative numbers. Note that the given scores, written in order from least to greatest, are −4, −3, and +1. It makes sense that the average, which is −2, lies between the least score, −4, and the greatest score, +1.

Lesson 15

Algebra

4-Quadrant Graphing

Students build upon their experiences with integers and graphing points in the first quadrant to graph points in the coordinate plane. They become familiar with all four quadrants by locating and plotting points and giving directions from one point to the next. These skills are useful in coordinate geometry and graphing equations.

Try It! Perform the Try It! activity on the next page.

Objective

Graph points in the coordinate plane.

Skills

- Identifying and plotting points
- Using a coordinate system
- Reasoning

NCTM Expectations

Grades 6–8
Algebra
- Model and solve contextualized problems using various representations, such as graphs, tables, and equations.

Grades 3–5
Geometry
- Make and use coordinate systems to specify locations and to describe paths.

Talk About It

Discuss the Try It! activity.

- **Ask:** *When you use an ordered pair to describe a position, which coordinate is related to the east and west directions? To the north and south directions?*
- **Ask:** *Why does Quadrant II have a negative x-coordinate and a positive y-coordinate?*
- **Ask:** *If you make Alison's house the origin, what are the coordinates of the school?*

Solve It

Reread the problem with students. Have students write the directions, using ordered pairs, from school to Alison's house and then from Alison's house to the movie theater and then from the movie theater to Janet's house.

More Ideas

For other ways to teach graphing points in the coordinate plane—

- Make a series of steps for students to follow, such as *place a centimeter cube on (–3, 0), move 4 units right and 5 units down; then move 2 units right and 8 units up*. For each step, have students place a cube and write the coordinates for the point to which they are directed.

- Extend the lesson by having students graph various landmarks around the school, such as buildings, parks, and offices. Give coordinates and have students locate the landmark with a centimeter cube. Then have students write instructions on how to move from one landmark to another.

Standardized Practice

Have students try the following problem.

Which point is at (–3, –2)?

A. Point Q   **C.** Point S

B. Point R   **D.** Point T

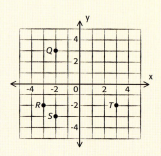

Try It! 20 minutes | Pairs

Here is a problem about graphing points in the coordinate plane.

Janet is going to Alison's house after school. Then they are going to the movie theater. Alison uses a grid to make a map for Janet. School is located at the origin. Each unit is one block. Directions east of the school are positive x-coordinates, north are positive y-coordinates, west are negative x-coordinates, and south are negative y-coordinates. Alison's house is 4 blocks east and 3 blocks south of school. Janet's house is 2 blocks west and 7 blocks south of school. The theater is 7 blocks west and 9 blocks north of Alison's house. What does the map look like?

Introduce the problem. Then have students do the activity to solve the problem. Distribute centimeter cubes, graph paper, and pencils. Draw a coordinate plane on the board with labels for the x- and y-axes. Label each quadrant with its number and sign pair, e.g., Quadrant II (−, +).

Materials
- centimeter cubes (10 per pair)
- 4-Quadrant Graph Paper (BLM 14; 1 per pair)
- pencils (1 per pair)

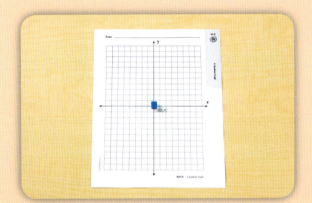

1. Say: *Look at the coordinate plane on the board.* **Ask:** *What is the origin of the coordinate plane?* Guide students to plot the coordinates on their graphs and to place a cube at (0, 0).

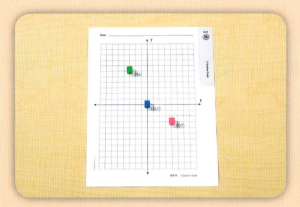

2. Have students locate and place a cube at Alison's house (4, -3) and at the movie theater (-3, 6). **Ask:** *What quadrant is Alison's house in? The movie theater?* **Say:** *Write the ordered pair describing each location.*

⚠ Look Out!

If students are confused about the directions of Alison's home and the movie theater, have them write *North, South, East,* and *West* next to the axes on their graphs. Explain that any location northeast of school is in Quadrant I, any location northwest is in Quadrant II, any location southwest is in Quadrant III, and any location southeast is in Quadrant IV.

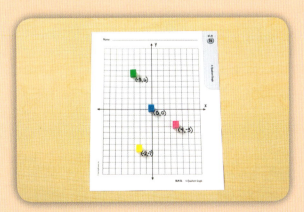

3. Have students locate and place a cube at Janet's house (-2, -7). **Ask:** *What quadrant is Janet's house in?* **Say:** *Write the ordered pair describing the location of Janet's house.*

Lesson 16

Algebra
Graphing Linear Equations

Knowledge about graphing ordered pairs enables students to graph equations. Students generate ordered pairs for an equation by substituting several *x* values into the equation and determining the corresponding values of *y*. They graph the equation by plotting the ordered pairs as points on a grid. If the equation is a linear equation, the points will lie on a straight line.

Objective
Graph linear equations on a four-quadrant grid.

Skills
- Using operations with integers
- Evaluating expressions
- Plotting ordered pairs on a coordinate plane

NCTM Expectations
Grades 6–8
Algebra
- Represent, analyze, and generalize a variety of patterns with tables, graphs, words, and, when possible, symbolic rules.
- Identify functions as linear or nonlinear and contrast their properties from tables, graphs, or equations.
- Model and solve contextualized problems using various representations, such as graphs, tables, and equations.
- Use graphs to analyze the nature of changes in quantities in linear relationships.

Try It! *Perform the Try It! activity on the next page.*

Talk About It
Discuss the Try It! activity.
- **Ask:** *When choosing values of x to substitute into an equation, why is it a good idea to choose at least one positive value, one negative value, and zero?*
- **Ask:** *Is it easier to see if (-3, -4) is a solution to $y = 2x + 2$ by using the graph or by using the table?* Have students explain their responses.

Solve It
Reread the problem with students. Have students explain in writing how to graph an equation. Then ask them to write a paragraph describing the graph of one of the equations, focusing on how the graph changes, the direction of the graph, where the graph intersects each axis, and so on.

More Ideas
For other ways to teach graphing linear equations—
- Have groups of students use centimeter cubes and grid paper to graph one of these equations: $y = 3x$, $y = x \div 3$, $y = x + 3$, and $y = x - 3$. Have one student transfer each graph to a separate transparency. Overlay the four graphs on an overhead projector and have students compare the graphs. Explain that equations whose graphs are straight lines are called *linear equations*.
- Show students a graph of a linear equation. Have them choose a least five coordinates from the graph to create a function table. Ask students to write the equation for the function table and graph.

Standardized Practice
Have students try the following problem.

Which ordered pair is a solution to the linear equation $y = x + 6$?

A. (1, 6)   **B.** (1, 7)   **C.** (6, 1)   **D.** (7, 1)

Try It! 25 minutes | Groups of 4

Here is a problem about graphing linear equations.

Two hikers are finding their way to camp. There are two trails they could follow. One trail can be mapped on a four-quadrant grid using the equation $y = 2x + 2$. The other trail can be mapped using the equation $y = x + 3$. Camp is located at the origin. Which trail will get the hikers closer to camp?

Introduce the problem. Then have students do the activity to solve the problem. Distribute centimeter cubes, graph paper, Function Tables, rulers, and pencils. Have students label the *x*-axis *East* on the right and *West* on the left and the *y*-axis *North* at the top and *South* at the bottom.

Materials
- centimeter cubes (15 per group)
- 4-Quadrant Graph Paper (BLM 14; 2 per group)
- Function Tables (BLM 10; 1 per group)
- rulers (1 per group)
- pencils (1 per student)

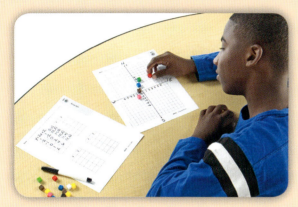

1. Have students write $y = 2x + 2$ at the top of the first table and fill in the *x* column with the values -2, -1, 0, 1, and 2. **Say:** *Substitute each value of x into the equation and find the corresponding value of y. Record the value. Then use the two values to write an ordered pair next to the table. Plot the ordered pairs on the grid using cubes.*

2. Ask: *Do the cubes appear to form a straight line?* Explain that the graph for a *linear equation* is a straight line. Have students replace each centimeter cube with a point and draw a line that passes through the points. Ask them to label the line $y = 2x + 2$.

⚠ Look Out!

Some students might confuse the *x*- and *y*-coordinates or the *x*- and *y*- axes. Remind them that the ordered pair is arranged in alphabetical order so the *x*-coordinate is listed first.

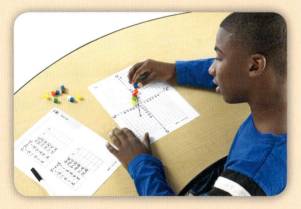

3. Guide students through the process once again as they graph $y = x + 3$. Have students describe each graph and then compare the two graphs. **Ask:** *Does either graph pass through the origin? Is there any point that is on both graphs?*

Measurement

Measurement is the assignment of a value to an attribute (length, area, or mass, for example) of an object. In the elementary grades, students first learn to determine what a measurable attribute is. Next, they become familiar with the units and processes associated with measuring. Finally, they begin to develop proficiency with the tools, techniques, and formulas used in measurement.

Students in grades 5 and 6 can begin to master higher-level skills such as estimating quantities other than length, selecting appropriate units, and choosing the needed degree of precision. They expand their understanding of measurement by investigating the perimeters and areas of two-dimensional figures and the volumes of three-dimensional figures. By comparing and contrasting the measurement of perimeter, area, and volume, students deepen their understanding of the formulas for these attributes. Students in these grades should begin to extend their knowledge of formulas beyond those for rectangles and rectangular solids to include triangles, circles, parallelograms, non-rectangular prisms, pyramids, and cylinders.

> **The NCTM Standards for Measurement suggest that students should:**
> - Understand measurable attributes of objects and the units, systems, and processes of measurement
> - Apply appropriate techniques, tools, and formulas to determine measurements

The study of measurement virtually requires the use of manipulatives and other concrete materials. It would be nearly impossible for students to gain a full understanding of measurement without handling objects, making physical comparisons, and using measuring tools. Manipulative activities that involve the measurement of derived quantities, such as area and volume, help students develop estimation skills that they can use to verify the reasonableness of computed answers. Square tiles and geoboards can be used to measure or estimate areas; cubes can be used to quantify volume. The following activities are built around manipulatives that students can use to develop skills and explore concepts in **Measurement**.

Measurement

Contents

Lesson 1   Standard Units and Precision 128
Objective: Use and compare standard units for measuring length.
Manipulatives: centimeter cubes and color tiles

Lesson 2   Perimeter and Area 130
Objective: Contrast the perimeter and area of a rectangle.
Manipulative: color tiles

Lesson 3   Area of a Parallelogram 132
Objective: Compare the areas of parallelograms and rectangles.
Manipulatives: geoboards

Lesson 4   Area of a Triangle 134
Objective: Compare the areas of triangles and parallelograms.
Manipulative: geoboards

Lesson 5   Surface Area of a Rectangular Solid 136
Objective: Find the surface area of a rectangular solid.
Manipulative: Snap Cubes®

Lesson 6   Volume of a Rectangular Solid 138
Objective: Find the volume of a rectangular solid.
Manipulative: Snap Cubes®

Lesson 7   Volumes of Prisms and Pyramids ... 140
Objective: Find the volumes of prisms and pyramids.
Manipulative: Relational GeoSolids®

Lesson 8   Circumference of a Circle and π ... 142
Objective: Find the circumference of a circle.
Manipulative: Relational GeoSolids®

Lesson 9   Area of a Circle 144
Objective: Find the area of a circle.
Manipulative: fraction circles

Lesson 1

Measurement
Standard Units and Precision

The precision of a measurement depends on the unit used. In this activity, students gain experiences that will help them understand precision. By using centimeter cubes to measure length to the nearest centimeter, students decide, for example, whether a segment is closer to 14 centimeters or 15 centimeters. Likewise, students can measure length to the nearest inch using color tiles. One method is more precise than the other.

Try It! *Perform the Try It! activity on the next page.*

Objective
Use and compare standard units for measuring length.

Skills
- Measuring lengths
- Using customary and metric units
- Understanding precision

NCTM Expectations

Grades 3–5
Measurement
- Understand the need for measuring with standard units and become familiar with standard units in the customary and metric systems.
- Understand that measurements are approximations and how differences in units affect precision.
- Select and apply appropriate standard units and tools to measure length, area, volume, weight, time, temperature, and the size of angles.

Talk About It
Discuss the Try It! activity.

- **Ask:** *If someone measures a line to the nearest whole tile and reports that the length is 6 tiles, in what range must the actual length be?* Since the length was estimated to the nearest whole tile, the length must be between five-and-a-half and six-and-a-half tiles.

- **Ask:** *If someone measures a line to the nearest whole cube and reports that the length is 15 cubes, in what range must the length be?* Since the length was estimated to the nearest whole cube, the length must be between fourteen-and-a-half and fifteen-and-a-half cubes.

Solve It
Reread the problem with students. Discuss why it is significant that the students' heights are reported as 53 inches and 132 centimeters, and not for example, as 53-and-zero sixteenths inches and 132-and-zero-tenths centimeters. Point out that because of the way the values are reported, we must assume that Katie is claiming her height to the nearest inch and that Ann is claiming her height to the nearest centimeter. Have students decide which claim is more precise and write an explanation of their reasoning.

More Ideas
For other ways to teach about standard units and precision—

- Have students estimate the lengths of the lines to the nearest half of a color tile and nearest half of a centimeter cube. Discuss the precisions of these measurements.

- Have students use a ruler to draw a line of specific length. Then ask them to draw additional lines that are shorter or longer but whose lengths could nonetheless be reported as the value that you specified. Repeat for a variety of lengths stated with different degrees of precision.

Standardized Practice
Have students try the following problem.

Which unit will give the most precise measurement?

A. foot   **B.** inch   **C.** meter   **D.** kilometer

Try It! 30 Minutes | Groups of 3

Here is a problem about standard units and precision.

Katie claims to be 53 inches tall. Her Canadian cousin Ann claims to be 132 centimeters tall. Which claim is more precise?

Introduce the problem. Then have students do the activity to solve the problem. Distribute centimeter cubes, color tiles, worksheets, rulers, and pencils to students. Explain that all measurements are approximations.

Materials
- centimeter cubes (50 per group)
- color tiles (20 per group)
- Line Measure worksheets (BLM 19; 1 per group)
- Customary/Metric rulers (1 per group)
- pencils (1 per group)

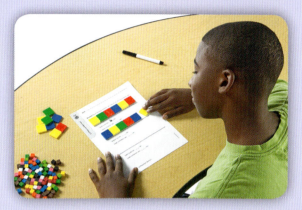

1. Have students use color tiles to measure the lengths of the three lines on the activity sheet and record each length to the nearest whole tile. **Ask:** *How do the measurements compare?*

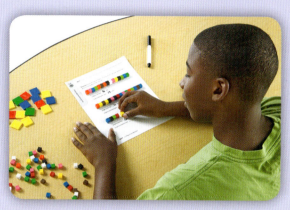

2. Have students use centimeter cubes to measure the lengths of the lines and record each length to the nearest whole cube. **Ask:** *How do the measurements compare?*

3. Say: *Explain why your three color-tile measurements are the same but your three centimeter-cube measurements are different.* Introduce the meaning of *precision* in measurement and how precision is related to the size of the measuring unit. Have students line up a row of centimeter cubes alongside a row of color tiles. **Ask:** *Which ruler measures more precisely—one made with tiles as units or one made with cubes as units?*

⚠ Look Out!

Make sure students understand that even though the second line on the activity sheet appears to be exactly 6 color tiles long, this does not mean that a measurement made with color tiles is more precise than a measurement made with centimeter cubes. Precision is related to what the reported measurement means to another person. If the reported value is 6 tiles, then another person knows that the length is between five-and-a-half and six-and-a-half tiles. The other person does not know that the length happens to be exactly 6 tiles.

Lesson 2

Measurement

Perimeter and Area

Perimeter is the distance around a figure, so it is a length measured in linear units such as inches. Area is a measure of the surface covered by a figure, so it is measured in square units such as square inches. Students can develop (or verify) formulas for determining the perimeter and area of a rectangle.

Objective

Contrast the perimeter and area of a rectangle.

Skills

- Building rectangles
- Measuring
- Differentiating between perimeter and area

NCTM Expectations

Grades 3–5
Measurement
- Understand such attributes as length, area, weight, volume, and size of angle and select the appropriate type of unit for measuring each attribute.
- Explore what happens to measurements of a two-dimensional shape such as its perimeter and area when the shape is changed in some way.
- Develop, understand, and use formulas to find the area of rectangles and related triangles and parallelograms.

Try It! Perform the Try It! activity on the next page.

Talk About It

Discuss the Try It! activity.

- **Ask:** *What is the perimeter of a rectangle with dimensions 5 units by 3 units?* Students can count units around the rectangle to get 16, or they can add: 5 + 3 + 5 + 3 = 16. **Say:** *Now write a formula for the perimeter of a rectangle with dimensions a units by b units.*
  Students write $P = a + a + b + b$, or $P = 2a + 2b$, or $P = 2(a + b)$.

- **Ask:** *What is the area of a rectangle with dimensions 5 units by 3 units?* Students can count square units that make up the rectangle to get 15, or they can multiply: 5 × 3 = 15. **Say:** *Now write a formula for the area of a rectangle with dimensions a units by b units.* Students write $A = ab$.

Solve It

Reread the problem with students. By completing the worksheet, students find all the possible 16-block routes for Lisa, starting with 7 blocks east, then 6, 5, 4, 3, 2, and 1. When comparing areas, students see that rectangles with the same perimeter can have different shapes and therefore different areas. Also, the rectangle with the greatest area is the square.

More Ideas

For other ways to teach about perimeter and area—

- Have students use centimeter cubes or color tiles to build all the possible rectangles with an area of 24 square units. Compare the perimeters.
  Ask: *Which rectangle has the least perimeter?*

- Students can use a geoboard and rubber bands for this activity.
  Step 1: Make a square with side length 2 units. Find its area and perimeter. (2-by-2 square, $A = 4$ and $P = 8$)
  Step 2: Make a rectangle that has the same area as your square, but different perimeter. (1-by-4 rectangle, $A = 4$ and $P = 10$)
  Step 3: Make a rectangle that has the same perimeter as your square, but different area. (1-by-3 rectangle, $A = 3$ and $P = 8$).

Standardized Practice

Have students try the following problem.

A rectangle has a perimeter of 28 feet. What is the greatest possible area?

A. 56 sq ft.   **B.** 49 sq ft.   **C.** 48 sq ft.   **D.** 40 sq ft.

Try It! 30 minutes | Groups of 4

Here is a problem about perimeter and area.

Each morning, Lisa walks 16 blocks for exercise. She always walks east, turns right and walks south, turns right and walks west, then turns right and walks north back to home. Find all the possible 16-block routes that Lisa can take. Compare the areas of the rectangles formed by the routes. Every block is the same length.

Introduce the problem. Then have students do the activity to solve the problem. Distribute color tiles, recording charts, and pencils. Have students fill in the recording chart with the following headings: *Dimensions, Perimeter, Area.*

Materials
- color tiles (100 per group)
- 4-Column Recording Chart (BLM 17; 1 per group)
- pencils (1 per group)

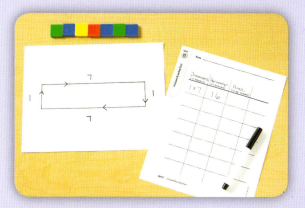

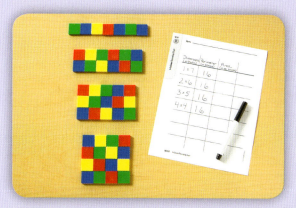

1. Have students line up 7 color tiles left to right. **Ask:** *What are the dimensions of the rectangle you made with the tiles?* Have students draw and label a copy of the rectangle. **Ask:** *What is the perimeter of the rectangle?* Have a student draw arrows on the diagram to show Lisa's route.

2. Say: *Use tiles to build a route in which Lisa first walks 6 blocks east.* Students form a 2-by-6 rectangle. **Say:** *Continue building rectangles that decrease by one block east each time.* Students form 3-by-5 and 4-by-4 rectangles. Have them fill in the dimensions and perimeter on the recording chart. Keep the models of the rectangles.

⚠ Look Out!

Students might think that since the length of the route is 16 blocks, they will use 16 tiles to build the rectangle that shows the route. This is not the case, because the route is found along the outside edge of the rectangle. Two sides of one tile are counted at each corner.

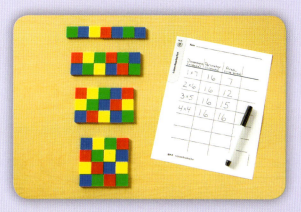

3. Say: *Now consider the areas of the rectangles that are enclosed by the routes. Add the area of each rectangle to the recording chart.*

Lesson 3

Measurement
Area of a Parallelogram

In this activity, students reason about the area of a parallelogram by visualizing its relationship to a corresponding rectangle. Without a formula, the area of a rectangle can be found by counting the number of square units that fill it. But since a parallelogram does not have right angles, it cannot be filled with whole squares. An area formula is needed. The area formula for parallelograms is a general form of the formula for rectangles.

Objective
Compare the areas of parallelograms and rectangles.

Skills
- Building figures
- Finding area
- Reasoning

NCTM Expectations
Grades 3–5
Measurement
- Develop, understand, and use formulas to find the area of rectangles and related triangles and parallelograms.

Try It! Perform the Try It! activity on the next page.

Talk About It
Discuss the Try It! activity.
- **Ask:** *How did the shapes of the parking spaces change on the geoboard when you shifted the bottoms to the right?*
- **Ask:** *How can you reason that the new shapes have the same area as the original shapes?*
- Explain how the area formula remains essentially the same except that the length of the rectangle is now the height of the parallelogram

Solve It
Reread the problem with students. Have students write a paragraph that describes how the area formula for a parallelogram is a general form of the area formula for a rectangle.

More Ideas
For other ways to teach about areas of parallelograms—
- Using the square and two small triangle tangrams, demonstrate how a rectangle can be made into a parallelogram having the same height and base length, and discuss why the areas are the same.
- Students can use 4 AngLegs™ to form a rectangle, then push on opposite corners to slant the rectangle and form a parallelogram. Ask students to compare the area of the rectangle with the area of the parallelogram. The area of the parallelogram is less, because the height has decreased.

Standardized Practice
Have students try the following problem.

A parallelogram has an area of 20 square inches and a base length of 4 inches. What is the height?

A. 4 inches

B. 5 inches

C. 6 inches

D. 10 inches

Try It! 20 Minutes | Pairs

Here is a problem about areas of parallelograms.

Doug wants to install three parking spaces on a rectangular section of land in front of his store. The section of land is 18 feet deep by 36 feet wide. Doug considers installing perpendicular spaces that are 9 feet wide. He also considers angled spaces. Does Doug's choice of layout affect the amount of paving that he needs for the parking spaces?

Introduce the problem. Then have students do the activity to solve the problem. Distribute geoboards, rubber bands, grid paper, and colored pencils to students.

Materials
- geoboard (1 per pair)
- rubber bands (6 per pair)
- 10 x 10 Grid Paper (BLM 3; 1 per pair)
- colored pencils (2 colors per pair)

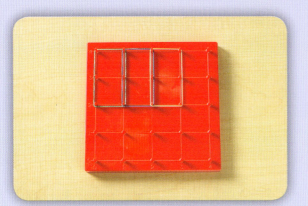

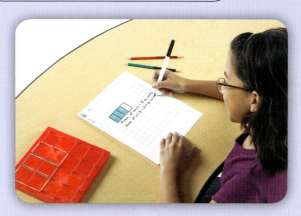

1. Say: *Using the upper half of a geoboard to represent the section of land, model three perpendicular parking spaces.* If necessary, guide students to realize that each space should be depicted using a 2-by-1 rectangle.

2. Say: *Draw a 2-by-4 rectangle on a grid to represent the section of land. Draw and shade three rectangles to represent perpendicular parking spaces.* **Ask:** *What is the area of each rectangle? What is the area of three rectangles added together?* Have students write their answers.

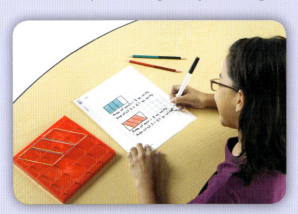

⚠ Look Out!

Point out to students that only one side length (called the base) is used to find the area of a parallelogram. The other measurement used is the height of the parallelogram, which is not a side length. Note that any of the four sides of a parallelogram can be called a base and the base is perpendicular to the height. The base and height are multiplied to find the area. With a rectangle, both the base and height (or length and width) are sides of the rectangle.

3. Say: *On the geoboard, shift the bottom of each rectangle to the right by one unit to create a model of three angled parking spaces.* **Ask:** *What is the area of each parallelogram and all three together?* Have students draw the model on their grid. Guide them to see that a triangular area has been uncovered on the left but that an equal area has been added on the right.

Lesson 4

Measurement
Area of a Triangle

The formula for the area of a parallelogram is $b \times h$, or *base times height*. Squares, rectangles, and parallelograms can all be divided in half to form congruent triangles. Any triangle made in this way has an area that is half the area of the original figure. Most students already have experience with this fact. For example, they might have seen a square sandwich cut in half to make two triangles. They can probably reason that each triangle has an area that is half the area of the whole sandwich. The same reasoning can be applied using a parallelogram as the original figure. This leads to the general formula for the area of a triangle: $A = \frac{1}{2} \times b \times h$.

Try It! Perform the Try It! activity on the next page.

Objective
Compare the areas of triangles and parallelograms.

Skills
- Building and drawing figures
- Finding area
- Comparing shapes

NCTM Expectations

Grades 3–5
Measurement
- Understand such attributes as length, area, weight, volume, and size of angle and select the appropriate type of unit for measuring each attribute.
- Develop, understand, and use formulas to find the area of rectangles and related triangles and parallelograms.

Talk About It
Discuss the Try It! activity.

- **Ask:** *How can you compare the area of the triangle with the area of the parallelogram without using formulas?*
- **Ask:** *If you know that the area of a parallelogram is $b \times h$, how do you reason that the area of a triangle is $\frac{1}{2} \times b \times h$?*
- **Ask:** *When is the height of a triangle equal to the length of a side?*

Solve It
Reread the problem with students. The area of the triangle is half the area of the related parallelogram. So the area of the triangle is 1 square unit and the area of the parallelogram is 2 square units.

More Ideas
For other ways to teach about the area of a triangle—

- Have students use pattern blocks to see the relationship between the area of a triangle and the area of a parallelogram. Two green triangles placed together cover the same area as one blue parallelogram.
- Have students use a geoboard to design a quilt block with various shapes of squares, rectangles, triangles, and parallelograms. Have them complete a table for the quilt block with the columns *Color, Shape, Area,* and *Total Area in Quilt Block*. Students can copy their quilt blocks onto grid paper and color them, then exchange blocks and analyze the areas of particular colors.

Standardized Practice
Have students try the following problem.

A parallelogram has a base of 7 feet and a height of 4 feet. What is the area of a triangle formed by drawing a diagonal on the parallelogram?

A. 7 square feet

B. 14 square feet

C. 28 square feet

D. 56 square feet

Try It! 30 Minutes | Groups of 4

Here is a problem about the area of a triangle.

Samira has made a triangle on a geoboard. Jo says she can find the area of the triangle by changing it into a parallelogram. How might Jo do this?

Introduce the problem. Then have students do the activity to solve the problem. Distribute geoboards, rubber bands, grid paper, paper, and pencils to students.

Materials
- geoboard (1 per group)
- rubber bands (2 per group)
- Centimeter Grid Paper (BLM 8; 1 sheet per group)
- paper (1 sheet per group)
- pencils (1 per group)

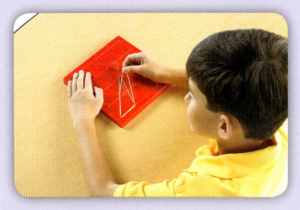

1. Say: *Make a triangle on the geoboard using the points (0, 0), (1, 0), and (2, 2).* Students use rubber bands to make the triangle. Have students place a second rubber band on top of the triangle.

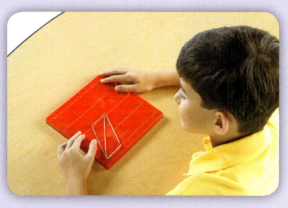

2. Say: *Now create a parallelogram with one of the rubber bands. Stretch the rubber band to create a fourth corner at the point (1, 2).* One of the rubber bands is stretched to the point (1, 2). **Say:** *Compare the triangle with the parallelogram.* Elicit that the triangle is half of the parallelogram.

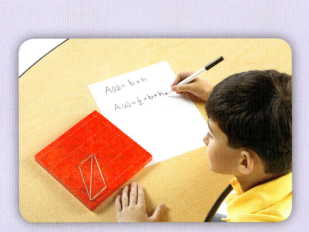

3. Ask: *What is the formula for finding the area of a parallelogram?* Elicit that the formula is $A(\square) = b \times h$. **Ask:** *What is the formula for finding the area of a triangle?* **Say:** *Use the models on the geoboard to help you determine the formula.* Elicit that the formula is $A(\triangle) = \frac{1}{2} \times b \times h$.

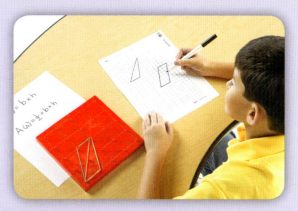

3. Say: *Draw the triangle and parallelogram on grid paper and compare their areas.* **Ask:** *What is the area of each shape?*

Lesson 5

Measurement

Surface Area of a Rectangular Solid

Using a model to explore and find surface area helps students distinguish surface area from volume and helps them visualize the dimensions of each face of a solid. In this lesson, students build a rectangular solid and use it to find surface area and to develop a formula for finding surface area.

Try It! Perform the Try It! activity on the next page.

Objective

Find the surface area of a rectangular solid.

Skills

- Finding area
- Measuring
- Reasoning

NCTM Expectations

Grades 6–8
Measurement
- Understand, select, and use units of appropriate size and type to measure angles, perimeter, area, surface area, and volume.
- Develop strategies to determine the surface area and volume of selected prisms, pyramids, and cylinders.

Talk About It

Discuss the Try It! activity.

- **Ask:** *How are rectangular solids alike? How can they be different?*
- **Ask:** *What does each square unit in the model represent?*
- **Say:** Pairs of faces on a rectangular solid are congruent. **Ask:** *How can you use this fact to help you find the total surface area?*
- **Ask:** *Why is a formula helpful for finding the surface area of a rectangular solid?*

Solve It

Reread the problem with students. Have them explain two methods of finding the surface area of a rectangular solid.

More Ideas

For other ways to teach about surface area of rectangular solids—

- Students work in pairs, using centimeter cubes to build two models of rectangular solids in which the dimensions of one are twice the dimensions of the other, such as $4 \times 6 \times 2$ and $2 \times 3 \times 1$. Have students find and compare the surface areas of the solids.
- Students work in pairs, using Snap Cubes® to build various models of rectangular solids that represent real-world situations. One student adds the areas of faces or counts square units to find surface area, while the other uses a formula. They compare solutions and switch roles.

Standardized Practice

Have students try the following problem.

How much construction paper does Bryant need to cover a pencil box that is 14 inches long, 6 inches wide, and 1 inch high?

A. 84 square inches

B. 188 square inches

C. 202 square inches

D. 208 square inches

Try It! 25 Minutes | Groups of 4

Here is a problem about the surface area of a rectangular solid.

Teri plans to cover all sides of a jewelry box in fabric. How much fabric does she need if the length of the jewelry box is 6 inches, the width is 4 inches, and the height is 2 inches? Write a formula for the surface area of a rectangular solid.

Introduce the problem. Then have students do the activity to solve the problem. Distribute Snap Cubes®, paper, and pencils to students. **Say:** The surface area of a rectangular solid is the sum of the areas of its faces.

Materials
- Snap Cubes® (50 per group)
- paper (1 sheet per group)
- pencils (1 per group)

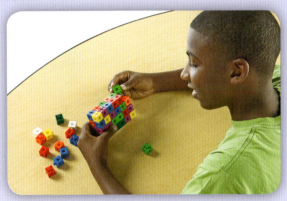

1. Say: *Use Snap Cubes to build a rectangular solid that is 6 units long, 4 units wide, and 2 units high.* **Ask:** *How many faces does the solid have? How many pairs of faces are congruent?* Students should see that there are six faces and three pairs of congruent faces.

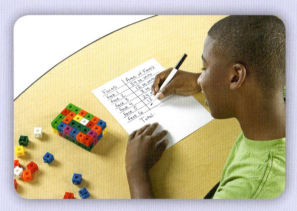

2. Ask: *How can you find the areas of the faces of the rectangular solid?* Guide students to see that they can count the number of square units on the faces or they can multiply the dimensions of the faces. **Say:** *Record the areas of the faces. Add the areas of the faces to get the total surface area.*

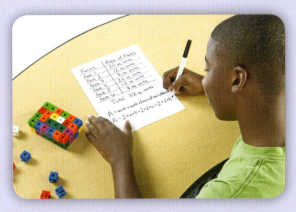

3. Say: *You can write a formula for surface area.* **Ask:** *Which dimensions do you multiply together for each pair of faces?* Encourage students to use their models to see that they multiply length by width, length by height, and width by height. Help them write the formula.

⚠ Look Out!

When calculating surface area, students may inadvertently add a pair of faces more than once, such as length × width, and at the same time, overlook a different pair of faces, such as length × height. Encourage students to record the dimensions or areas of each pair of faces so they can check for this error. This is especially important when the numbers are easy to compute mentally.

Lesson 6

Measurement
Volume of a Rectangular Solid

Students explore the volume of a rectangular solid by constructing a model out of cubic units. By counting units, students build an understanding of volume and visualize the connection between volume and dimensions. With this understanding, students can develop formulas for volume and are prepared to consider the volumes of other solids.

Try It! Perform the Try It! activity on the next page.

Objective
Find the volume of a rectangular solid.

Skills
- Finding volume
- Measuring
- Reasoning

NCTM Expectations
Grades 6–8
Measurement
- Understand, select, and use units of appropriate size and type to measure angles, perimeter, area, surface area, and volume.
- Select and apply techniques and tools to accurately find length, area, volume, and angle measures to appropriate levels of precision.
- Develop strategies to determine the surface area and volume of selected prisms, pyramids, and cylinders.

Talk About It
Discuss the Try It! activity.

- **Ask:** *How is a cubic unit different from a square unit?* Have students use a Snap Cube® to demonstrate the difference.
- **Ask:** *How can you determine the number of cubes in the rectangular solid by looking at the completed model?*
- **Say:** *You can calculate the volume of a rectangular solid by finding the area of the solid's base and multiplying this by the solid's height.* **Ask:** *How do you write the formula represented by this method?* Guide students to write $V = B \times h$. **Ask:** *How do you find the area of the base?* Students should know to multiply length by width.

Solve It
Reread the problem with students. Have them talk about the two ways of writing the formula for the volume of the freezer. Guide them to see how the formulas represent different ways of visualizing the same thing, and ask them to write a paragraph about what they have learned.

More Ideas
For other ways to teach about volume of rectangular solids—

- Give pairs of students 48 Snap Cubes or centimeter cubes and have them build five different rectangular solids with a volume of 48 cubic units. Have them record and compare the dimensions of the solids.
- Have students use the 2 cubes and the 2 rectangular prisms from the Relational GeoSolids® set to estimate the volumes of the solids and to investigate the relationship between volume and side lengths.

Standardized Practice
Have students try the following problem.

Construction workers dug a hole that measures 5 meters long by 4 meters wide by 3 meters deep. What is the volume of the hole?

A. 12 cubic meters   C. 60 cubic meters
B. 27 cubic meters   D. 94 cubic meters

Try It! 30 Minutes | Groups of 3

Here is a problem about the volume of a rectangular solid.

Mr. Adams bought a freezer. The freezer is rectangular and the space inside it measures 3 feet long by 2 feet wide by 5 feet high. What is the volume of the space inside the freezer? Write a general formula for this volume that could be used for any length, width, and height.

Introduce the problem. Then have students do the activity to solve the problem. Distribute Snap Cubes®, paper, and pencils to students. **Say:** Volume is a three-dimensional measure and is therefore expressed in cubic units.

Materials
- Snap Cubes® (30 per group)
- paper (1 sheet per group)
- pencils (1 per group)

1. Say: Using cubes, build a 3-by-2 rectangle to represent the bottom layer of space inside the freezer. **Ask:** How many cubic units are in this layer of cubes? Have students count cubes or multiply to find the number of cubic units.

2. Say: Stack additional 3-by-2 layers on top of the first layer to make a solid that is five layers tall. The solid model represents the space inside the freezer. **Ask:** How many cubic units are in the model? Guide students to multiply the number of cubic units in each layer by the number of layers. Have them record their results.

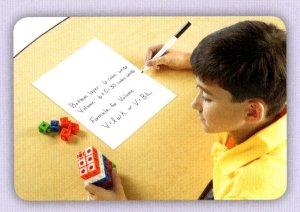

3. Say: You can multiply length by width by height to find the volume. Have students confirm that multiplying length by width by height produces the same answer. **Say:** Write a formula for volume. Help students write the formula.

⚠ Look Out!

Some students might not see the connection between the different forms of the volume formula: $V = lwh$ and $V = Bh$. Help students see that multiplying length by width is the same as calculating the area of the base. Reinforce this idea by presenting a table like the following and having students extend the table for additional layers.

Number of layers	Cubic units	$V = lwh$	$V = Bh$
one	6	$3 \times 2 \times 1 = 6$	$6 \times 1 = 6$
two	12	$3 \times 2 \times 2 = 12$	$6 \times 2 = 12$

Measurement

Lesson 7

Measurement

Volumes of Prisms and Pyramids

Students use models to develop formulas for the volumes of prisms and pyramids. They compare the fill capacities of the models and use these comparisons to justify the formulas for the volumes—prisms: $V = B \times H$ and pyramids: $V = \frac{1}{3} \times A \times H$ where A is the area of the base and H is the height of the solid.

Try It! Perform the Try It! activity on the next page.

Objective

Find the volumes of prisms and pyramids.

Skills

- Multiplying
- Finding area
- Reasoning

NCTM Expectations

Grades 6–8
Measurement
- Understand both metric and customary systems of measurement.
- Understand, select, and use units of appropriate size and type to measure angles, perimeter, area, surface area, and volume.
- Select and apply techniques and tools to accurately find length, area, volume, and angle measures to appropriate levels of precision.
- Develop strategies to determine the surface area and volume of selected prisms, pyramids, and cylinders.

Talk About It

Discuss the Try It! activity.

- Have students compare the faces of the four GeoSolids®. Encourage them to identify corresponding faces among the shapes, such as the triangle bases of the triangular prism and the triangular pyramid.

- Discuss with students the meanings of B, b, H, and h in the volume formulas. For example, tell students that b is the base length of the prism's triangular base and that h is the altitude of this triangle.

- **Ask:** *How does the volume of a square pyramid change if one of its dimensions is doubled? If two of its dimensions are doubled? If three of its dimensions are doubled?*

Solve It

Reread the problem with students. Have students sketch the four ornament shapes and explain how to find each volume.

More Ideas

For other ways to teach about volumes of prisms and pyramids—

- Have students draw and cut out the partial nets for a cube and rectangular prism on the Centimeter Grid (BLM 8). Have students tape the sides and fill each figure with centimeter cubes. Students should count the cubes and compare this number to the volume found using the formula $V = B \times H$ for each figure.

- Have students use centimeter cubes to build models of the Relational GeoSolids® and estimate the volume of each. Then introduce the formulas.

Standardized Practice

Have students try the following problem.

Each side of the base of a square pyramid is 12 feet. The height of the pyramid is 10 feet. What is the volume?

A. 1440 cu ft   **B.** 480 cu ft   **C.** 360 cu ft   **D.** 40 cu ft

Try It! 30 minutes | Groups of 4

Here is a problem about volumes of prisms and pyramids.

Anika is filling containers with sand to make yard ornaments. How can she calculate the volume of sand needed to fill an ornament shaped like a cube? A square pyramid? A triangular prism? A triangular pyramid?

Introduce the problem. Then have students do the activity to solve the problem. Distribute the Relational GeoSolids®, fill material, rulers, paper, and pencils to students.

Materials
- Relational GeoSolids® (large cube, large triangular prism, triangular pyramid, square pyramid; 1 set per group)
- sand, rice, or other substance to fill GeoSolids (2 cups per group)
- centimeter rulers (1 per group)
- paper (1 sheet per group)
- pencils (1 per group)

1. Have students determine how many square pyramids of sand it takes to fill the cube. **Ask:** *What is the relationship between the volumes of the cube and the square pyramid?*

2. Write *V (cube)* = *B* × *H* on the board. Remind students that *B* is the area of the base and have them calculate the volume of the cube using lengths measured to the nearest centimeter. **Say:** *Use the volume of the cube to determine the volume of the square pyramid.* Introduce the formula *V*(☐ pyramid) = $\frac{1}{3}$ × *B* × *H*.

3. Have students determine how many triangular pyramids of sand it takes to fill the triangular prism. **Ask:** *What is the relationship between the volumes of the triangular prism and triangular pyramid?*

4. Write the formulas *V*(△ prism) = *B* × *H* and *B* (△ prism) = $\frac{1}{2}$ × *b* × *h* on the board. Have them calculate the volume of the triangular prism. **Say:** *Find and use the volume of the triangular prism to determine the volume of the triangular pyramid.* Introduce the formula *V* (△ pyramid) = $\frac{1}{3}$ × ($\frac{1}{2}$ × *b* × *h*) × *H*.

Lesson 8

Measurement

Circumference of a Circle and π

Students look at the ratio of circumference to diameter for various circles and develop both an approximation of the value of π and the formula for finding circumference. While a cingle circle shows the ratio, a larger number of examples helps students recognize the consistency of the ratio and provides a stronger basis for making a generalization.

Try It! Perform the Try It! activity on the next page.

Objective

Find the circumference of a circle.

Skills

- Measuring
- Generalizing
- Multiplying and dividing

NCTM Expectations

Grades 3–5
Measurement
- Understand both metric and customary systems of measurement.
- Select and apply techniques and tools to accurately find length, area, volume, and angle measures to appropriate levels of precision.
- Develop and use formulas to determine the circumference of circles and the area of triangles, parallelograms, trapezoids, and circles and develop strategies to find the area of more-complex shapes.

Talk About It

Discuss the Try It! activity.

- **Ask:** *When might it be more useful to use $\frac{22}{7}$ as an approximation for π? When might it be more useful to use 3.14 for π?*
- **Ask:** *How would you find the circumference of a circle if you know its radius?* Explain that since the diameter is twice the length of the radius, the value 2r can be substituted for d in the formula for finding circumference: $C = πd = 2πr$.
- **Ask:** *How can you find the diameter of a circle if you know its circumference?*

Solve It

Reread the problem with students. Have students explain how to find the circumference of a circle when the diameter is known. Then have them find the length of ribbon Kaden needs to fit exactly around the top edge of the can.

More Ideas

For other ways to teach about circumference and π —

- Have students trace the inner and outer circles of the fraction circle rings, and then measure the circumferences of the traced circles to develop the concept of π.
- Provide each group with a different circular object, such as a two-color counter, spinner, fraction circle, or Relational GeoSolids®. Have each group find the ratio of circumference to diameter of their object. Record results on the board and have students generalize the ratio—that is, determine π.

Standardized Practice

Have students try the following problem.

The diameter of a circle is 52 inches. Which expression can you evaluate to find the circumference?

A. $52 ÷ π$ B. $π ÷ 52$ C. $52 × π$ D. $26 × π$

142

Try It!
30 minutes | Groups of 4

Here is a problem about finding the circumference.

Kaden is decorating a can for his mother to store her small crafts. He wants to glue a piece of ribbon to the top edge of the can so that it goes around the can exactly one time. How much ribbon does he need if the diameter of the can is 14 cm?

Introduce the problem. Then have students do the activity to solve the problem. Distribute the large and the small cylinders, other circular objects, recording chart, paper, pencils, string, rulers, and calculators. Have students start a recording chart with these headings: *Object, Diameter (d), Circumference (C),* and $\frac{C}{d}$.

Materials
- Relational GeoSolids® large and small cylinder (1 set per group)
- Other circular objects (optional)
- 4-Column Recording Chart (BLM 17; 1 per group)
- paper (1 sheet per group)
- pencils (1 per group)
- string (2 feet length per group)
- centimeter rulers (1 per group)
- calculators (1 per group)

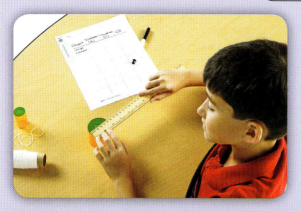

1. Have students measure the diameter and circumference of the base of the large and small cylinders and record each measurement to the nearest tenth of a centimeter. Then have students divide to complete the chart.

2. Have students measure the diameter and circumference of other circular objects to the nearest tenth of a centimeter and complete the table. **Ask:** *What pattern do you see in the measurements?* Write the symbol π on the board. **Say:** *This symbol is called pi. We often use 3.14 or $\frac{22}{7}$ to approximate its value.*

⚠ Look Out!

Be sure that students measure each diameter and circumference correctly. Remind them to measure the diameter at the widest part of the circle. This will help students calculate a more accurate number for π. Explain to students that π is the same for any circle, no matter how big or small. Students' calculations for π may differ slightly.

3. Ask: *How can you find the circumference of a circle if you know its diameter? What formula can you use?* Write $C = \pi \times d$ on the board. **Say:** *Add a circle with a diameter of 14 cm to your recording sheet. Use the formula to find the circumference.*

Lesson 9

Measurement

Area of a Circle

Measurement concepts are closely related to other mathematics topics, such as geometry and algebra. To develop and conceptualize the formula for the area of a circle, students first estimate the area by tracing a circle on grid paper. Moving the parts of the circle to form a shape that resembles the more-familiar parallelogram helps students justify and internalize the formula.

Objective
Find the area of a circle.

Skills
- Measuring
- Estimating area
- Reasoning

NCTM Expectations

Grades 6–8
Measurement
- Understand both metric and customary systems of measurement.
- Select and apply techniques and tools to accurately find length, area, volume, and angle measures to appropriate levels of precision.
- Develop and use formulas to determine the circumference of circles and the area of triangles, parallelograms, trapezoids, and circles and develop strategies to find the area of more-complex shapes.

Try It! *Perform the Try It! activity on the next page.*

Talk About It

Discuss the Try It! activity.

- **Ask:** *Why is the area of the circle written in square centimeters?*
- **Ask:** *What is the relationship between the radius, diameter, circumference, and area of a circle?*
- **Ask:** *How would you find the area of a circle if you know its diameter?*

Solve It

Reread the problem with students. Have students list the information that is needed to find the area of a circle. Then have them explain how to find the area of Maya's rug.

More Ideas

For other ways to teach about the area of a circle—

- Have students use centimeter cubes to estimate the area of a circle. Then, using the cubes, have students estimate the radius and diameter of the circle to calculate the area. Tell students to compare the two methods for finding the area of the circle.

- Have students find the area of the circular base of a solid from a set of Relational GeoSolids®. Have students calculate the area two ways using 3.14 and $\frac{22}{7}$ for π.

Standardized Practice

Have students try the following problem.

The radius of a circle is 10 mi. What is the area of the circle to the nearest whole number?

A. 63 mi

B. 100 sq mi

C. 314 mi

D. 314 sq mi

Try It! 20 minutes | Pairs

Here is a problem about the area of a circle.

Maya has a circular rug in her bedroom. What is the area of the rug if the radius is 4.4 feet?

Introduce the problem. Then have students do the activity to solve the problem. Distribute the fraction circles, grid paper, paper, pencils, and calculators to students. Review the terms *radius* and *diameter*. Write the symbol π on the board. Have students give the approximate value of π as a fraction and as a decimal.

Materials
- fraction circles (1 set per pair)
- Centimeter Grid Paper (BLM 8, 1 per pair)
- paper (1 sheet per pair)
- calculators (1 per pair)
- pencils (1 per pair)

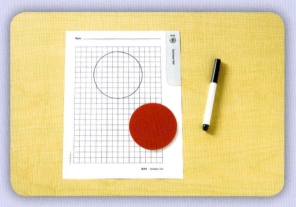

1. Have students trace the red circle on the grid paper. **Say:** Estimate the area of the circle by counting the squares and parts of squares. Have students share their estimates.

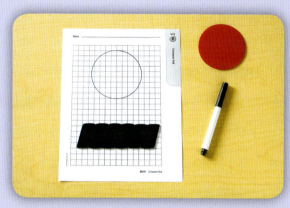

2. Guide students to arrange the 12 twelfths in a side-by-side pattern on the grid paper. **Ask:** *What shape does your arrangement resemble?* Write $A = b \times h$ on the board. Have students explain how to find the area of a parallelogram and have them estimate the base, height, and area of the figure.

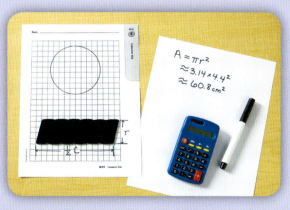

3. Ask: *What part of a circle is shown by the base of the arrangement? The height?* Show students that the base of the *parallelogram* is roughly $\frac{1}{2}C$ and that the height is roughly r. Write the area of the *parallelogram* $A = \frac{1}{2}C \times r$ on the board. Replace C with $2\pi r$ and simplify to get the formula for the area of the circle, $A = \pi r^2$. Have students calculate the area of the red circle using radius 4.4 cm.

⚠ Look Out!

Some students may confuse the radius and the diameter of a circle. Have them draw and label the parts of a circle. Point out that area is always measured in square units, even when the shape has curved sides. Watch for students who think that r^2 means to multiply the length of the radius by two. Review the meaning of exponents with these students.

Data Analysis and Probability

In **Data Analysis and Probability** students learn to collect, organize, process, and display data to answer questions. They also learn about the notion of chance and apply it to the study of random events. In time, students will make use of these skills and concepts in real-world situations; for example, to interpret information in the media and to become intelligent consumers.

The NCTM Standards for Data Analysis and Probability suggest that students should:

- Formulate questions that can be addressed with data and collect, organize, and display relevant data to answer them
- Select and use appropriate statistical methods to analyze data
- Develop and evaluate inferences and predictions that are based on data
- Understand and apply basic concepts of probability

At grades 5 and 6, students become more adept at discriminating between the various representations of data, such as bar graphs, line graphs, circle graphs, and scatterplots, and they learn to choose the representation that will best reflect the answer to a given question. They learn that a graph, being a visual representation, or picture, of a data set, can reveal trends and suggest relationships among quantities. Students will also deepen their understanding of mean, median, mode, and range, and will learn how to select the best measure to use for a given situation.

Probability instruction in these grades focuses on making and testing conjectures. Students learn to distinguish between theoretical and experimental probability. They expand their understanding of the notion of chance as they explore the probabilities of single and compound events, and they begin to understand more deeply the connection between the notion of chance and the calculation of probability. The following activities are built around manipulatives that students can use to develop skills and explore concepts in **Data Analysis and Probability.**

Data Analysis and Probability
Contents

Lesson 1 Mean, Median, Mode, and Range . . 148
Objective: Find and compare the mean, median, mode, and range of a data set.
Manipulative: centimeter cubes

Lesson 2 Make a Conjecture Using a Scatterplot 150
Objective: Create and use a scatterplot to make a conjecture.
Manipulative: centimeter cubes

Lesson 3 Line Graphs 152
Objective: Display data in a line graph.
Manipulative: centimeter cubes

Lesson 4 Circle Graphs 154
Objective: Display data in a circle graph.
Manipulative: fraction circles

Lesson 5 Counting Principle 156
Objective: Use the Counting Principle to find the number of possible outcomes of a compound event.
Manipulative: two-color counters and centimeter cubes

Lesson 6 Probability of an Event 158
Objective: Find the theoretical and experimental probabilities of an event.
Manipulative: spinners

Lesson 7 Complementary and Mutually Exclusive Events 160
Objective: Understand complementary and mutually exclusive events.
Manipulative: color tiles

Lesson 8 Probability of a Compound Event 162
Objective: Find the probability of a compound event.
Manipulative: polyhedra dice and two-color counters

Lesson 1

Data Analysis and Probability

Mean, Median, Mode, and Range

Students might be familiar with the meaning of average. Here they learn that the average, or mean, is just one of the measures of center that can be ascribed to a set of data. The median and mode are also measures of center. The range is a measure of spread.

Try It! Perform the Try It! activity on the next page.

Objective
Find and compare the mean, median, mode, and range of a data set.

Skills
- Organizing data
- Comparing data
- Interpreting data

NCTM Expectations
Grades 6–8
Data Analysis and Probability
- Find, use, and interpret measures of center and spread, including mean and interquartile range.

Talk About It
Discuss the Try It! activity.

- Talk about how a set of data can have more than one mode or no mode.
- **Ask:** *Why is it important to order the data first?* **Say:** *Sometimes there are two middle values, so you have to find the mean, or average, of the two middle values to find the median.* **Ask:** *How can you tell that a data set has two middle values?*
- **Ask:** *Which measures are most likely to change if you remove the least and greatest values from the data set? What are some situations in which you would want to know the mean, median, mode, or range?*

Solve It
Reread the problem with students. Ask them to describe each measure and how to find it. Then have them compare measures and explain their answers.

More Ideas
For other ways to teach mean, median, mode, and range—

- Provide data sets with six values such as 9, 6, 1, 10, 9, and 7. Demonstrate how to find the median when there are two middle values. Have students work in pairs using color tiles and/or two-color counters to model the data and find the mean (7), median (8), mode (9), and range (9). Have students discuss how each measure represents the data.
- Give students a partial-data set of values such as 4, 9, 7. Tell students the data set has five values total. The mean and median is 6, the mode is 4, and the range is 5. Have students use color tiles to find that the data set is 4, 4, 6, 7, and 9. Give students other partial-data sets and measures.

Standardized Practice
Have students try the following problem.

Ryan sent the following numbers of e-mails each day last week: 4, 12, 10, 6, 3, 12, and 9. Find the mean, median, mode, and range of the data set. Which measure is the greatest?

A. mean   B. median   C. mode   D. range

Try It! 20 minutes | Pairs

Here is a problem about the mean, median, mode, and range of a data set.

A pet store has the following numbers of fish in seven different aquariums: 6, 12, 7, 5, 11, 6, and 9. Compare the mean, median, mode, and range of the fish in the aquariums. The pet store owner doesn't want any aquarium to have too many more fish than another. Which data measure would he be interested in?

Introduce the problem. Then have students do the activity to solve the problem. Distribute centimeter cubes, grid paper, paper, and pencils. Define mean, median, mode, and range.

Materials
- centimeter cubes (60 per pair)
- Centimeter Grid Paper (BLM 8; 1 per pair)
- paper (1 sheet per pair)
- pencils (1 per pair)

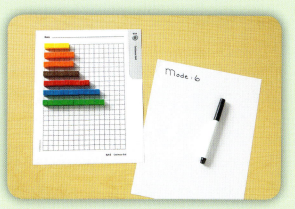

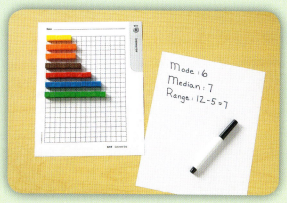

1. Have students use centimeter cubes to show the number of fish in each aquarium and arrange the cubes on the grid in rows using one color of cube for each value. Then have them order the values from least to greatest. **Say:** *The mode is the value that occurs most often.* **Ask:** *What is the mode of this data set?*

2. Say: *The median is the middle value.* **Ask:** *What is the median of this data set?* **Say:** *The range is the difference between the greatest value and the least value.* **Ask:** *What is the range of this data set?*

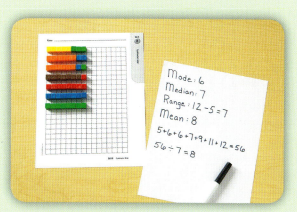

3. Have students arrange the cubes so that there is an equal number in each row. Explain how this number is the mean, or average. **Say:** *You can also compute the mean. Add all of the values in the data set and divide the sum by the total number of data values.* Help students compute the mean.

⚠ Look Out!

Watch for students who confuse median and mean. Emphasize that when the data items are listed in order according to size, the median is the middle number. The mean is the average. Have students arrange two sets of cubes for the data set 3, 5, 5, 8, and 9, with one set of cubes arranged to determine the median and the other arranged to demonstrate the mean. Then have them compare and contrast the mean and median of the data set.

Data Analysis and Probability

Lesson 2

Data Analysis and Probability
Make a Conjecture Using a Scatterplot

Students apply the skills of organizing and representing data to examine relationships between characteristics in a population. Scatterplots are used to test for trends in data— that is, to test for correlations between characteristics. The lesson prepares students to explore other relationships in or between populations.

Objective
Create and use a scatterplot to make a conjecture.

Skills
- Representing data
- Making conjectures
- Analyzing data

NCTM Expectations
Grades 6–8
Data Analysis and Probability
- Make conjectures about possible relationships between two characteristics of a sample on the basis of scatterplots of the data and approximate lines of fit.

Try It! *Perform the Try It! activity on the next page.*

Talk About It
Discuss the Try It! activity.

- **Ask:** *Will Keenan's data be appropriate for all students in the class? Have students explain their answers.*
- **Ask:** *Why might more than one point have the same x-coordinate? The same y-coordinate?*
- Discuss the positive trend shown by the scatterplot. **Ask:** *How would a scatterplot show a negative trend? How might it show no trend?*

Solve It
Reread the problem with students. Have them discuss the trend in the data. Ask students to make a conjecture about how quiz scores depend on hours of study. Stress, however, that a trend does not prove a cause-and-effect relationship.

More Ideas
For other ways to teach about scatterplots—

- Extend this problem by having students create tables of values from their lines of best fit. Have them use their tables to predict scores based on number of hours of study.

- Encourage students to use real data, such as baseball or basketball statistics from their local newspaper, to create a scatterplot. Have students use centimeter cubes to plot the data on a grid. Then have them pencil in the data points and draw a line of best fit.

Standardized Practice
Have students try the following problem.

Which scatterplot shows a negative trend?

A. B. C. D.

Try It! 20 minutes | Pairs

Here is a problem about making and using a scatterplot.

Keenan wants to see if there is a relationship between the number of hours a classmate studies and the number of questions he or she gets correct on a quiz. Keenan uses a sample of ten classmates. Below is a table representing the sample. Is there a relationship between the hours of study and the number of questions answered correctly?

Hours of Study	3	6	1	4	3	3	5	4	5	6
Number of questions correct	4	7	2	5	3	2	5	3	4	5

Introduce the problem. Then have students do the activity to solve the problem. Distribute centimeter cubes, grid paper, rulers, and pencils to students and analyze that data.

Materials
- centimeter cubes (15 per pair)
- Centimeter Grid Paper (BLM 8; 1 per pair)
- pencil (1 per pair)
- ruler (1 per pair)

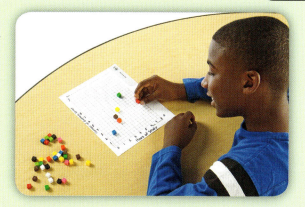

1. Ask students to set up the axes on their grids and plot the data from the table using centimeter cubes. **Say:** *Let the x-axis represent the hours of study and the y-axis represent the number of questions correct.*

2. Introduce the term *scatterplot*. **Say:** *A scatterplot is a plot of data points. If the data points rise from left to right, then they show a positive trend. If the data points fall from left to right, then they show a negative trend. If the data points neither rise nor fall, then they show no trend.* **Ask:** *Do you see a trend? What kind? What conjecture would you make from the data?* Elicit that the data show a positive trend between the number of hours of study and the number of questions answered correctly.

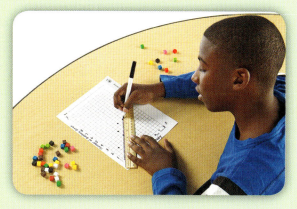

3. Have students pencil in the data points. Discuss how to draw a line of best fit. Students should draw a straight line such that there are about the same number of data points above and below the line and such that the points are about the same average distance from the line. **Ask:** *Which coordinates does your line pass through?*

⚠ Look Out!

Some students may connect the points to make a line graph. Increase the data size, if necessary, to demonstrate that it is not meaningful to connect all the points. Point out that the line of fit only helps to show the trend in the data — nothing more.

Lesson 3

Data Analysis and Probability

Line Graphs

Line graphs typically display changes in data over time. They can show trends such as increases or decreases. Ordered pairs of data are plotted, then the plotted points are connected with line segments. The segments approximate continuous change that occurs between data points.

Try It! Perform the Try It! activity on the next page.

Objective
Display data in a line graph.

Skills
- Plotting points
- Interpreting graphs
- Making predictions

NCTM Expectations

Grades 3–5
Data Analysis and Probability
- Represent data using tables and graphs such as line plots, bar graphs, and line graphs.
- Propose and justify conclusions and predictions that are based on data and design studies to further investigate the conclusions or predictions.

Talk About It

Discuss the Try It! activity.

- **Ask:** *How many data points are given? How many years are covered?*
- **Ask:** *Is it easier to see the shape of the data when it is graphed as five points, or after a line is formed?*
- **Ask:** *Can your line graph be used to predict Grant's height when he is 10 years old? 20 years old? Explain your answer.*

Solve It

Reread the problem with students. Have students identify when Grant's height increased the most and when it increased the least. A reasonable prediction for Grant's height at 5 years old is 41 to 42 inches.

More Ideas

For other ways to teach about line graphs—

- Have students collect data, such as, the daily high temperature for one week, or the population of your state for the past ten years. Graph the data on centimeter grid paper. Talk about the best way to set up the graph and what labels and increments to use. Have students use centimeter cubes to model the data on centimeter grid paper and then draw a line graph and make predictions.

- Have students use two-color counters to generate data, then draw a line graph to display the data. Put 40 counters in a bag. Shake the bag and spill the counters onto a desk. Count the number of red counters. Mark the point for the number of red counters for trials 1 through 10. Connect the points with a line segment and give the graph a title. Have students look for a trend in the data and discuss what they see.

Standardized Practice

Have students try the following problem.

Using the line graph, find the best estimate of Grant's height when he was $2\frac{1}{2}$ years old.

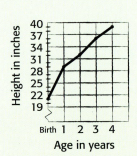

A. 30 inches
B. 32 inches
C. 34 inches
D. 36 inches

Try It! 25 minutes | Pairs

Here is a problem about line graphs.

The table shows Grant's height, measured to the nearest inch, for four years of growth. Display the data in a line graph. In which year did Grant's height increase the most? In which years did it increase the least? Predict Grant's height at 5 years old.

Age in years	Birth	1	2	3	4
Height in inches	21	29	32	36	39

Introduce the problem. Then have students do the activity to solve the problem. Distribute centimeter cubes, grid paper, and pencils. Have students set up a graph with "Age in years" on the horizontal axis and "Height in inches" on the vertical axis.

Materials
- centimeter cubes (30 per pair)
- Centimeter Grid Paper (BLM 8; 1 per pair)
- pencils (1 per pair)
- colored pencils (1 per pair)

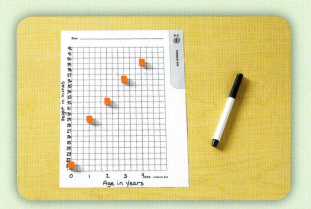

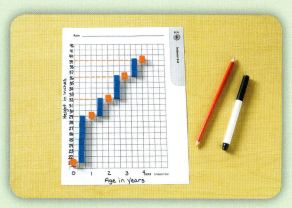

1. Say: *On the horizontal axis, let each year be represented by 3 units. On the vertical axis, let each inch be represented by 1 unit.* Have students show a break on the vertical axis between 0 and 21. **Say:** *Model each data point with a centimeter cube.* Students place cubes at (0, 21), (1, 29), (2, 32), (3, 36), and (4, 39).

2. Have students draw a horizontal line from each data point to the y-axis. **Say:** *Use cubes to measure the vertical distances between the five points.* **Ask:** *Which two data points have the most cubes between them? Which have the least?* Note that more cubes between data points indicate a greater increase in height.

⚠ Look Out!

Make sure students understand the difference between points in a line graph and those in a scatterplot. In both cases, ordered pairs are graphed in a coordinate plane. The points are connected in a line graph, but not in a scatterplot. Explain that the points are connected in a line graph because the data represent one subject over a period of time. A scatterplot may include data for several different subjects. The points do not represent a single, continuous change, so they are not connected.

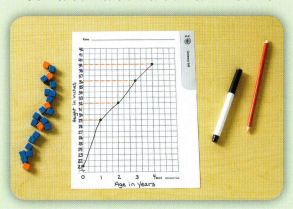

3. Say: *Now draw the line graph. Remove the cubes, mark the data points, and connect the points.* Students might want to refer back to the table to graph the points. Tell students to add a title to their graph. **Ask:** *What is your prediction for Grant's height at 5 years old?*

Lesson 4

Data Analysis and Probability

Circle Graphs

Some data are best presented in a circle graph. A circle graph is an area model that displays data as parts of a whole. The skills that students bring to this activity involve fractions, percents, circles, and angle measures.

Try It! Perform the Try It! activity on the next page.

Objective
Display data in a circle graph.

Skills
- Making graphs
- Writing fractions and percents
- Measuring angles

NCTM Expectations
Grades 6–8
Data Analysis and Probability
- Select, create, and use appropriate graphical representations of data, including histograms, box plots, or scatterplots.

Talk About It
Discuss the Try It! activity.

- **Ask:** *Why is it useful to write the fractions in simplest form?*
- **Ask:** *Why should the fraction circle parts form a complete circle? If they do not form a complete circle, what might the error be?*
- **Ask:** *Why is it useful to write the fractions as percents in your circle graph? What is the total of the percents?*
- **Ask:** *How can you find the angle measures for the parts of a circle graph if you have no fraction circles to model the parts?*

Solve It
Reread the problem with students. There are 48 vehicles on the lot — 50% cars, 16.6% trucks, 25% SUVs, and 8.3% vans. The sum of these percentages is 100%. The center of the circle is the vertex of the angles that define the different parts of the graph. The sum of these is 360°; a full circle. Students compare the fraction pieces and shade the cars section of the graph to show that it is the largest percent.

More Ideas
For other ways to teach about circle graphs—

- Give students additional sets of data to model. They can use fraction circles and rings, or a compass and protractor to make circle graphs.
- Students can use centimeter cubes or color tiles to generate data and then draw a circle graph to display the data. For example, grab a handful of cubes. Separate them by color, and then count the number of each color and find the total. For each color, write a fraction and multiply it by 360° to find the angle measure for the circle graph. Draw the circle graph and label it with colors and percents.

Standardized Practice
Have students try the following problem.

A class has 7 girls and 9 boys. Which circle graph depicts the data?

A.   B.   C.   D.

154

Try It! 20 minutes | Groups of 4

Here is a problem about circle graphs.

A used car lot has 24 cars, 8 trucks, 12 SUVs, and 4 vans for sale. Display the data in a circle graph. Label the graph with the percent of each type of vehicle on the lot. Which type of vehicle is represented by the largest percent?

Introduce the problem. Then have students do the activity to solve the problem. Distribute fraction circles, fraction circle rings, paper, and pencils.

Materials
- fraction circles (1 set per group)
- fraction circle rings (1 set per group)
- paper (1 sheet per group)
- pencils (1 per group)

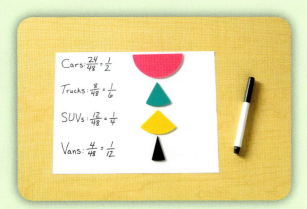

1. Say: *Write a fraction in simplest form to represent the ratio of each type of vehicle to the total number of vehicles. Model each fraction with one of the fraction parts.* Guide students to use a pink half, an aqua sixth, a yellow fourth, and a black twelfth. **Say:** *Place the four fraction parts together to form a circle.*

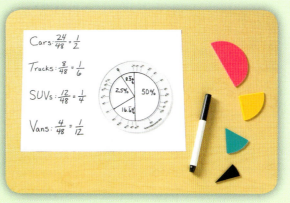

2. Say: *Use the percent ring to draw a circle.* Have students divide the circle into sections by tracing each fraction part. Place the ring so that zero is at the first edge of one of the sections. Guide students to read the percent on the ring at the other edge. **Say:** *Write a percent for the part.* Have students continue until they have written the percent for each part.

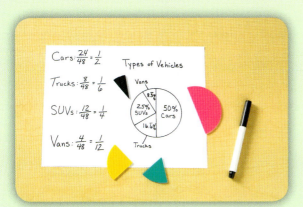

3. Say: *Label your graph with a title and the names of the types of vehicles.* Have students refer to the fraction pieces to determine the label for each section.

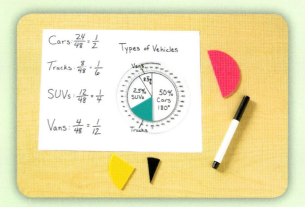

4. Have students put each fraction part inside the degree fraction circle ring. Have students determine the degree measure for each part. Be sure students realize that the total of the measures of the central angles in a circle is 360°.

Data Analysis and Probability

155

Lesson 5

Data Analysis and Probability

Counting Principle

In this lesson, students use the Counting Principle to find the number of possible outcomes of a compound event. Knowledge of the principle prepares students to learn about probability.

Try It! Perform the Try It! activity on the next page.

Objective

Use the Counting Principle to find the number of possible outcomes of a compound event.

Skills

- Analyzing probability situations
- Predicting probability outcomes
- Multiplying

NCTM Expectations

Grades 6–8
Data Analysis and Probability
- Compute probabilities for simple compound events, using such methods as organized lists, tree diagrams, and area models.

Talk About It

Discuss the Try It! activity.

- **Ask:** *How many single events are in the compound event?*
- **Ask:** *What are the possible outcomes when you toss a coin? Why do you multiply by two for each additional coin that you toss?*
- **Ask:** *How does the tree diagram help you verify the Counting Principle?*

Solve It

Reread the problem with students. Have students compare the methods (tree diagram and Counting Principle) of finding the number of possible outcomes. Then have them explain how to use the Counting Principle to find the number of possible outcomes for any compound event.

More Ideas

For other ways to teach the Counting Principle—

- Provide compound events involving spinners, number cubes, coins, and sacks of color tiles or centimeter cubes with 2 to 5 single events in the compound event. Have students model and find the possible outcomes of the compound event.

- Have student pairs use color tiles or centimeter cubes to model and find combinations involving real-world situations, such as combinations of hats, gloves, and coats, or various menu items. Name items so that students can determine the number of choices, such as red, blue, and brown hats, or wool, fleece, and leather gloves.

Standardized Practice

Have students try the following problem.

Diego tossed a six-sided number cube and spun a spinner with equal sections A through H. How many possible outcomes are there?

A. 12

B. 14

C. 36

D. 48

Try It! 20 minutes | Pairs

Here is a problem about the Counting Principle.

How many possible outcomes are there if Ava tosses two coins and draws a cube from a bag that contains 1 blue, 1 green, and 1 red centimeter cube?

Introduce the problem. Then have students do the activity to solve the problem. Distribute two-color counters, centimeter cubes, bags, paper, and pencils. Explain that an *event* is the *result* or *outcome* of an experiment, such as tails on a coin toss, and that a *compound event* is two or more events combined.

Materials
- two-color counters (6 per pair)
- centimeter cubes (5 each of blue, green, and red per pair)
- plastic or paper bags (1 per pair)
- paper (1 sheet per pair)
- pencils (1 per pair)

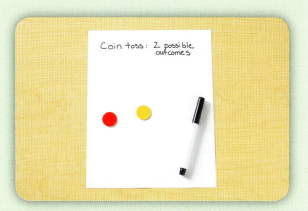

1. Ask: *How many outcomes are possible when you toss a coin?* Guide students to see that there are two possible outcomes since a coin has two different sides. Have students record the number.

2. Ask: *How many outcomes are possible for drawing a cube from the bag?* Guide students to see that there are three possible outcomes since there are three different cubes. Have students record the number.

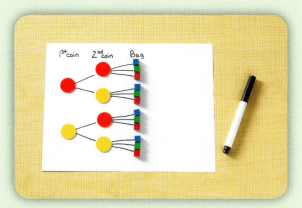

3. Have students build a tree diagram to show the possible outcomes for the compound event. Students model the possible outcomes of the first coin toss: red and yellow. For each of these, students model the possible outcomes of the second coin toss. Students complete the tree by modeling the possible outcomes of drawing a cube from the bag: blue, green, and red.

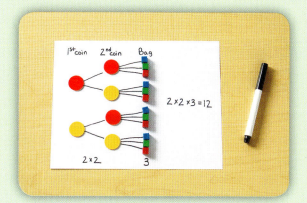

4. Guide students to find the number of possible outcomes of the compound event by multiplying the numbers for the simple events: $2 \times 2 \times 3$. Using the tree, have students verify their answer by counting the number of possible paths from left to right. Each path represents an outcome. **Ask:** *How many paths did you find?*

Data Analysis and Probability

Probability of an Event

Students learn that they can use theoretical probability to predict the results of an experiment and that it may or may not be the same as the experimental probability. Here they use spinners to learn how to distinguish and compare theoretical and experimental probability, and how to express the probability as a fraction, decimal, or percent.

Try It! Perform the Try It! activity on the next page.

Objective
Find the theoretical and experimental probabilities of an event.

Skills
- Analyzing
- Comparing
- Multiplying
- Determining equivalent fractions, decimals, and percents

NCTM Expectations
Grades 6–8
Data Analysis and Probability
- Use proportionality and a basic understanding of probability to make and test conjectures about the results of experiments and simulations.

Talk About It
Discuss the Try It! activity.

- **Ask:** *What does* trial *mean in the case of spinning a spinner?*
- Discuss that the likelihood that an event will occur is indicated by a number from 0 to 1. Zero means the event is impossible, and 1 means the event is certain.
- **Ask:** *Given the theoretical probability of the spinner landing on a number less than 4, how many times would you expect the spinner to land on one of these numbers in 10 spins? 30 spins? 50 spins?*
- **Ask:** *As you run more trials, what do you notice about the theoretical and experimental probabilities?*
- **Ask:** *Who do you think will earn more points, Thom or Maya? Why?*

Solve It
Reread the problem with students. Ask them to find the theoretical and experimental probabilities of Thom getting a point and to use their experiment to compare and contrast theoretical and experimental probability.

More Ideas
For other ways to teach about the probability of an event—

- Give polyhedra dice to pairs of students. Have one student find the experimental probability of rolling a prime number on 10, 20, and 30 rolls while the other finds and uses the theoretical probability to determine the expected results.
- Have pairs of students use color tiles to find the theoretical probability of drawing a certain color of tile from a bag. Have them predict the number of times the color will be drawn in 50 or 100 trials, and then have them find the experimental probability and compare it with their prediction.

Standardized Practice
Have students try the following problem.

A spinner is numbered 1 through 12. Lauren spins it 15 times and it lands on a number greater than 3 ten times. What is the theoretical probability that the spinner lands on a number greater than 3?

A. $\frac{1}{5}$ B. $\frac{1}{4}$ C. $\frac{3}{4}$ D. $\frac{4}{5}$

Try It! 30 minutes | Pairs

Here is a problem about the probability of an event.

Thom and Maya are playing a game with a spinner numbered 1–8. Thom gets a point if the spinner lands on a number less than 4. Maya gets a point if the spinner lands on 4 or greater. Compare the theoretical probability that Thom will get a point with the experimental probability using 10, 30, and 50 spins.

Introduce the problem. Then have students do the activity to solve the problem. Distribute spinners, paper, and pencils. Explain that theoretical probability describes what should occur, while experimental probability describes what actually occurs in an experiment.

Materials
- spinners (1 of the spinners from the set per pair)
- paper (1 sheet per pair)
- pencils (1 per pair)

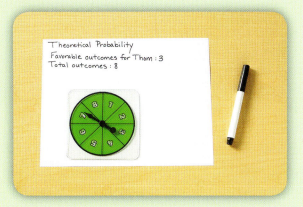

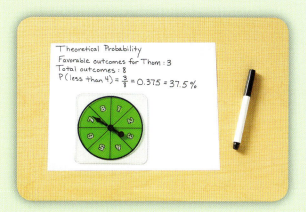

1. Say: *The theoretical probability of an event is the ratio of the number of favorable outcomes to the total number of possible outcomes.* **Ask:** *How many favorable outcomes are there for Thom? How many possible outcomes are there?*

2. Say: *You can express probability as a fraction, decimal, or percent.* Help students express as a fraction the theoretical probability of the spinner landing on a number less than 4. Then have them convert the fraction to a decimal and a percent.

3. Say: *Experimental probability is the ratio of favorable trials to the total number of trials.* Have students tally favorable trials (numbers less than 4) for 10, 30, and 50 spins, and express the experimental probabilities in three ways. Then have them compare the theoretical and experimental probabilities.

⚠ Look Out!

Some students may confuse theoretical and experimental probability. Emphasize that theoretical probability tells what would happen if each possible outcome appears the same number of times. For example, if a spinner has five equal sections and you spin it five times, each number would appear once, or 1 out of 5 times. In an actual experiment, a number might appear more than once in five spins. What actually happens is experimental probability.

Data Analysis and Probability

Lesson 7

Data Analysis and Probability

Complementary and Mutually Exclusive Events

Once students have a basic grasp of probability concepts, they can extend their understanding to include more than a simple event. They discover that probabilities include the possibility that an event occurs and the possibility that an event does not occur, as well as the possibility that two or more events cannot occur together.

Try It! Perform the Try It! activity on the next page.

Objective
Understand complementary and mutually exclusive events.

Skills
- Reasoning
- Analyzing
- Adding and subtracting fractions

NCTM Expectations

Grades 6–8
Data Analysis and Probability
- Understand and use appropriate terminology to describe complementary and mutually exclusive events.

Talk About It
Discuss the Try It! activity.

- **Ask:** *Why do complementary probabilities add up to 1?* Have students name examples of complementary events.
- Have students explain why picking a blue duck and picking a green duck are mutually exclusive events.
- **Ask:** *Are complementary events mutually exclusive? Explain. Are mutually exclusive events always complementary? Explain.*

Solve It
Reread the problem with students. Have them do the activities with the color tiles to compare and contrast complementary and mutually exclusive events, and explain how to find their probabilities.

More Ideas
For other ways to teach about complementary and mutually exclusive events—

- Give spinners to pairs of students. Each student formulates a probability question: one about complementary events, such as what is the probability of spinning an *8* and what is the probability of not spinning an *8*, and the other about mutually exclusive events. Students trade questions, find the probabilities by inspecting their spinners, and then discuss their results.
- Have pairs of students use one of the polyhedra die to determine the probabilities of complementary and mutually exclusive events. For example, students find the probabilities of rolling and not rolling a multiple of 3, and the probability of rolling a prime number or a 4.

Standardized Practice
Have students try the following problem.

A bag contains 3 blue tiles, 4 green tiles, 1 yellow tile, and 2 red tiles. What is the probability of drawing a green or red tile from the bag?

A. $\frac{1}{5}$ B. $\frac{2}{5}$ C. $\frac{1}{2}$ D. $\frac{3}{5}$

Try It! 25 minutes | Groups of 4

Here is a problem about complementary and mutually exclusive events.

Eve picks a prize from a bag filled with 3 blue ducks, 2 yellow ducks, 6 green ducks, and 1 red duck. What is the probability that Eve picks a yellow duck and what is the probability that she does not pick a yellow duck? What is the probability that Eve picks either a blue duck or a green duck?

Introduce the problem. Then have students do the activity to solve the problem. Distribute color tiles, paper, and pencils to students.

Materials
- color tiles (6 blue, 4 yellow, 2 red, and 12 green per group)
- paper (1 sheet per group)
- pencils (1 per group)

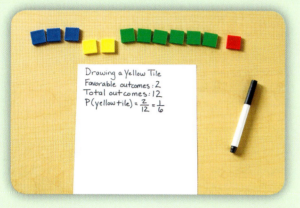

1. Say: Use color tiles to model the ducks in the bag. **Ask:** What is the probability of drawing a yellow tile? Have students model the possibilities and then count and record the favorable outcomes and the total outcomes. Have them express the probability as a fraction in simplest form.

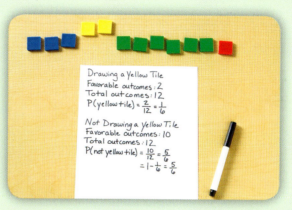

2. Say: The complement of drawing a yellow tile is **not** drawing a yellow tile. **Ask:** What is the probability of the complement? Have students show the non-yellow tiles. **Say:** Find and record the probability by listing and counting outcomes, and then by subtracting from 1 the probability of drawing a yellow tile.

3. Say: Mutually exclusive events are two or more possible events that cannot occur at the same time. **Ask:** Are the events of drawing a blue tile and drawing a green tile mutually exclusive? Have students model the events to show that both cannot occur at the same time.

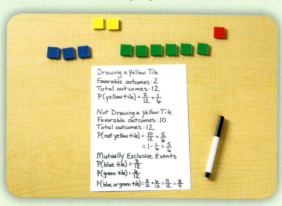

4. Ask: What is the probability of drawing either a blue or a green tile? Have students model the number of favorable selections—3 blue plus 6 green. Help them conclude that P (blue or green) is $\frac{9}{12}$, or $\frac{3}{4}$. Elicit, further, that this is the same as P(blue) + P(green)—that is, $\frac{3}{12} + \frac{6}{12} = \frac{9}{12} = \frac{3}{4}$.

Data Analysis and Probability

Lesson 8

Data Analysis and Probability

Probability of a Compound Event

Experience with experiments helps students build on their intuitive sense about probability. Making predictions about an event or events and then comparing the predictions with experimental data allows students to adjust their thinking and correct misconceptions. Students may use organized lists or tree diagrams to help them find probability. This lesson focuses on independent events in the form $P(A \text{ and } B) = P(A) \times P(B)$.

Objective
Find the probability of a compound event.

Skills
- Calculating probability
- Collecting, representing, and organizing data

NCTM Expectations
Grades 6–8
Data Analysis and Probability
- Use proportionality and a basic understanding of probability to make and test conjectures about the results of experiments and simulations.
- Compute probabilities for simple compound events, using such methods as organized lists, tree diagrams, and area models.

Try It! Perform the Try It! activity on the next page.

Talk About It
Discuss the Try It! activity.

- **Ask:** *What fraction do you use to compute the probability of an event?* Point out that the fraction is the number of favorable outcomes over the number of possible outcomes and that this applies to any probability.
- **Ask:** *Besides using an organized list, how else might you find the number of possible outcomes?*
- **Ask:** *How would the favorable outcomes differ if Deana wanted to roll a number greater than 5 **or** toss the color yellow?* Point out that students do not multiply to find this probability.

Solve It
Reread the problem with students. Have them find and write the probability for each single event in the form of a fraction. Guide students to multiply these fractions to find the probability of the compound event. Have students compare the fractions they calculated in Step 3.

More Ideas
For other ways to teach about probabilities of compound events—

- Have students conduct the experiment using a 1–8 spinner instead of the die.
- Use color tiles to conduct an experiment. Place six yellow and four red tiles in a bag. Have students draw one tile, record the color, replace the tile, and repeat. Each draw is a single event. The two draws are a combined event. Have students find the experimental and theoretical probabilities of drawing a yellow tile first and a red tile second.

Standardized Practice
Have students try the following problem.

Johann rolls a number cube with the numbers 1–6. He also tosses a coin. What is the probability that he will roll an even number and toss tails?

A. $\frac{1}{2}$ B. $\frac{1}{3}$ C. $\frac{1}{4}$ D. $\frac{1}{12}$

Try It! 30 minutes | Groups of 4

Here is a problem about the probability of a compound event.

Deana has a polyhedra die with faces labeled 1–8 and a two-color counter with one yellow and one red face. What is the probability that she will roll a number greater than 5 and toss a counter yellow-face up?

Introduce the problem. Then have students do the activity to solve the problem. Distribute dice, two-color counters, paper, and pencils to students. Explain that a compound event is a combination of two or more single events.

Materials
- polyhedra dice numbered 1–8 (1 per group)
- two-color counters (1 per group)
- cup (optional for rolling die and counters; 1 per group)
- paper (2 sheets per group)
- pencils (1 per group)

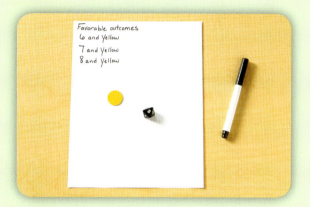

1. Ask: *What are the favorable outcomes for each single event? What are the favorable outcomes for the compound event?* Have students record the favorable outcomes for the compound event.

2. Guide students to list all the possible outcomes for the compound event. Then have them perform at least 50 trials for this event and record their results. **Ask:** *What is the experimental probability and the theoretical probability for the compound event? How do they compare?*

⚠ Look Out!

Emphasize that both events must have favorable outcomes to satisfy the conditions: rolling a 2 and yellow is not a favorable outcome because the outcome is not favorable for one of the two single events. Students may think that the theoretical probability and the experimental probability should be the same. Stress that the two results may not be the same.

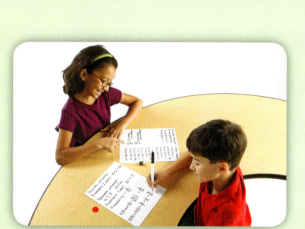

3. Guide students to see that they can determine the number of possible outcomes and the number of favorable outcomes by using the Counting Principle. Show how this leads to the rule $P(A \text{ and } B) = P(A) \times P(B)$.

BLM 1

Decimal Models

Name _____

= _____

+ _____

+ _____

Number: _____

BLM 1 Decimal Models

Name _____

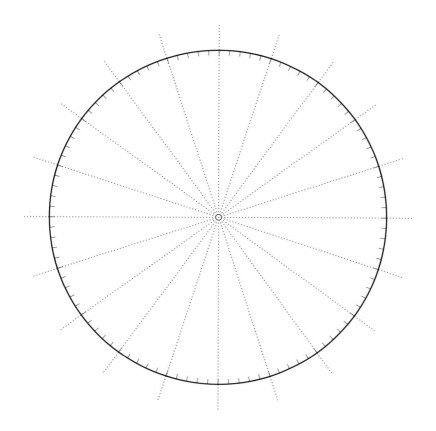

BLM 2 — Fraction Circle in Hundredths

	Percent	Fraction (Hundredths)	Decimal	Fraction (Reduced)
Basketball	20%			
Hockey				$\frac{1}{10}$
Soccer			0.25	
Other Sport				

Name _____

10 × 10 Grid

Name _____

Tile Arrays
BLM 4

Number of Tiles	Number of Arrays	Dimensions	Factors
1			
2			
3			
4			
5			
6			
7			
8			
9			
10			
11			
12			
13			
14			
15			
16			
17			
18			
19			

BLM 5

Hundred Chart

Name _____

1	2	3	4	5	6	7	8	9	10
11	12	13	14	15	16	17	18	19	20
21	22	23	24	25	26	27	28	29	30
31	32	33	34	35	36	37	38	39	40
41	42	43	44	45	46	47	48	49	50
51	52	53	54	55	56	57	58	59	60
61	62	63	64	65	66	67	68	69	70
71	72	73	74	75	76	77	78	79	80
81	82	83	84	85	86	87	88	89	90
91	92	93	94	95	96	97	98	99	100

Name

Inch Grid Paper

BLM 7 Eighths Fraction Squares

Name _____

$\dfrac{3}{4} \times \underline{} = \underline{}$

$\dfrac{3}{4} \div \underline{} = \underline{}$

Name

Centimeter Grid Paper

Name _____

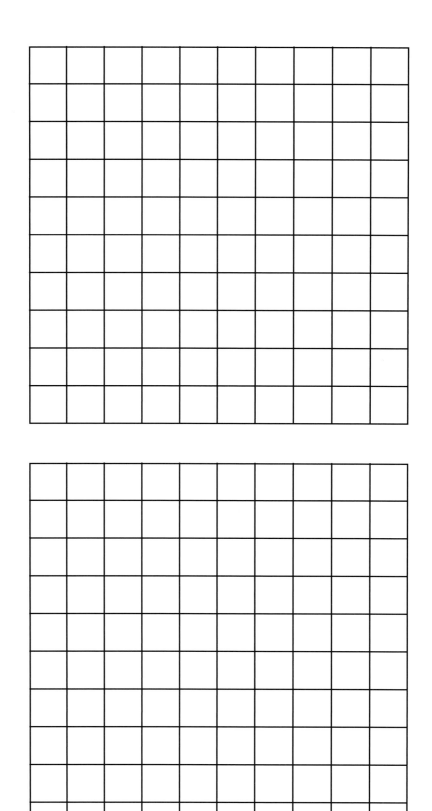

Hundredths Grids

BLM 9 Hundredths Grids

Name _____

x	y

x	y

x	y

x	y

BLM 10   Function Table

Name _____

BLM 11 1-inch Number Lines

Name _____

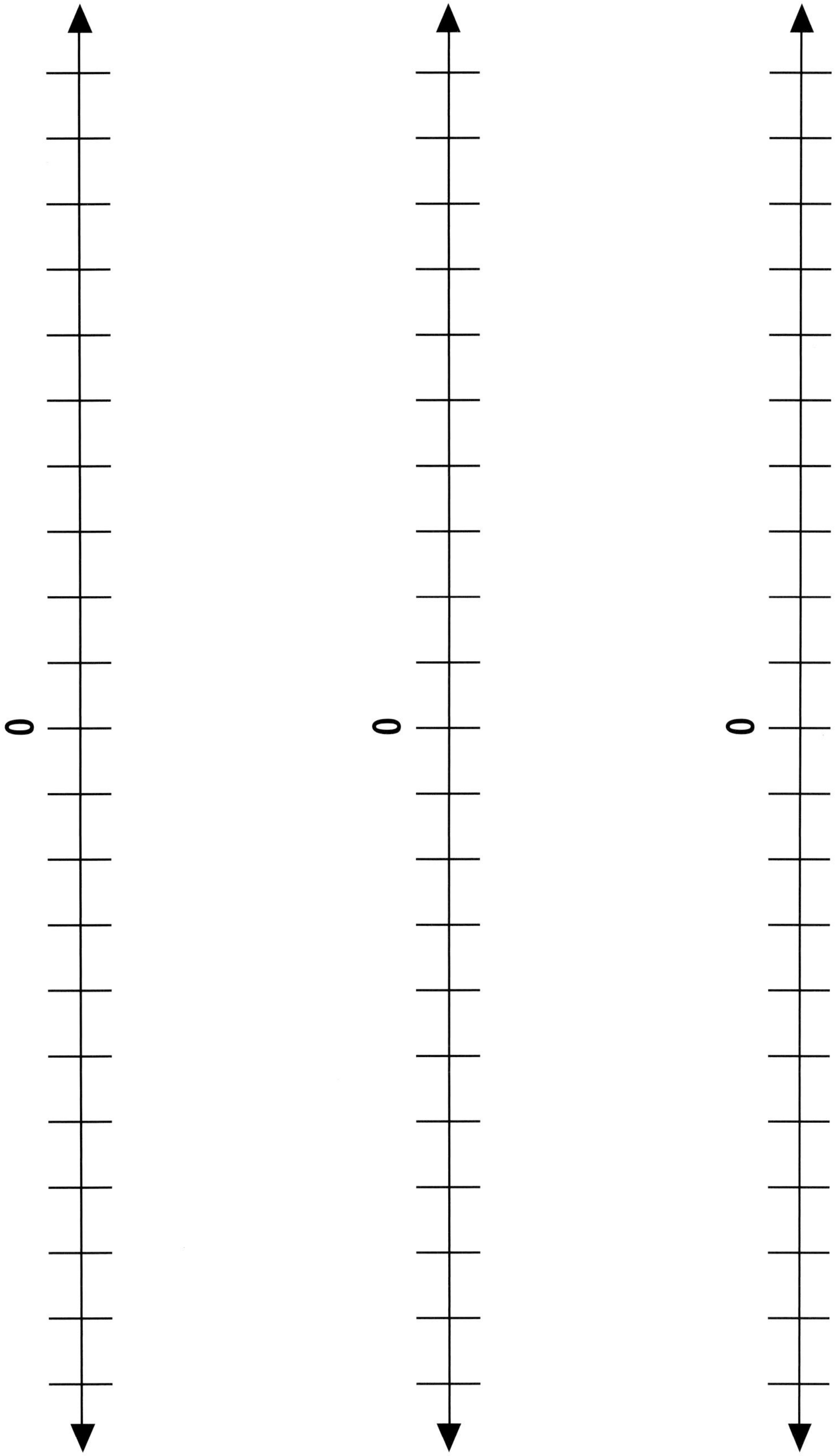

0   0   0

BLM 12   1-cm Number Lines

BLM 13

½-cm Number Lines

Name _____

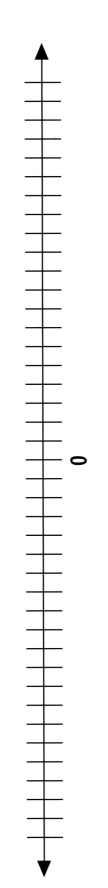 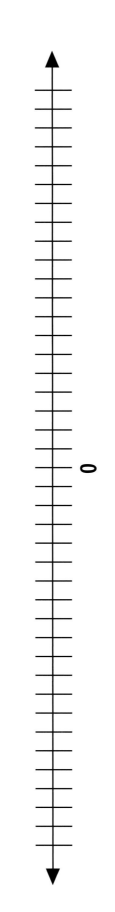 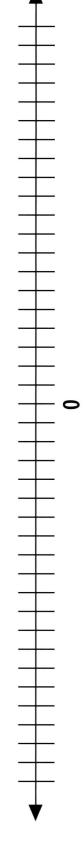

BLM 13 ½-cm Number Lines

Name _____

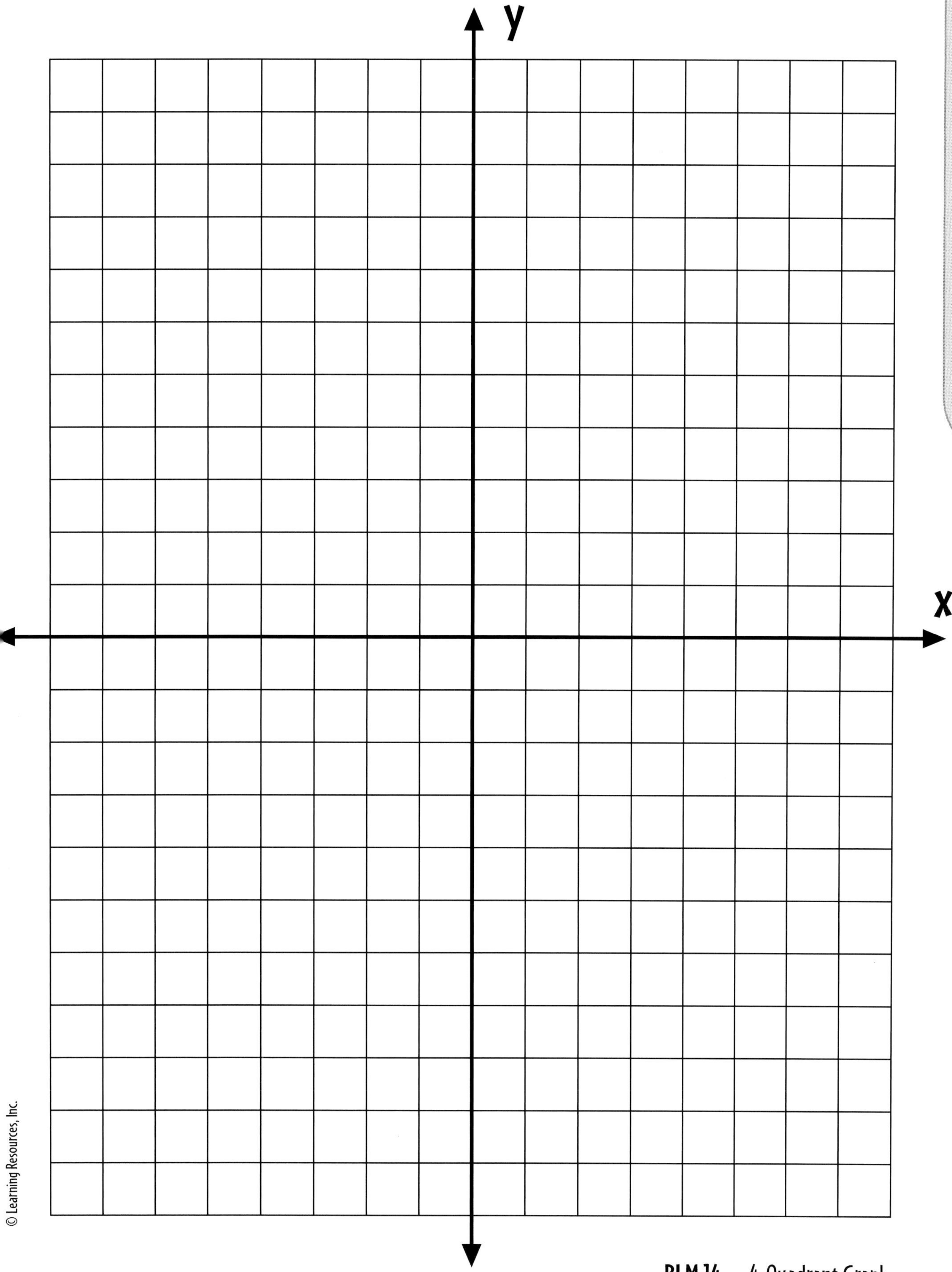

y

x

4-Quadrant Graph Paper

Name _____

Properties of Quadrilaterals

Name	Number of Pairs of Parallel Sides	Number of Congruent Sides	Number of Right Angles
Trapezoid	1	0, 2, or 3	0 or 2
Parallelogram	2	2	0 or 4
Rhombus	2	4	0 or 4
Square	2	4	4
Rectangle	2	2	4
Kite	0 or 2	2 or 4	0 or 4

Can you make the shape?

Name	Yes	No
Trapezoid		
Parallelogram		
Rhombus		
Square		
Rectangle		
Kite		

Name _____

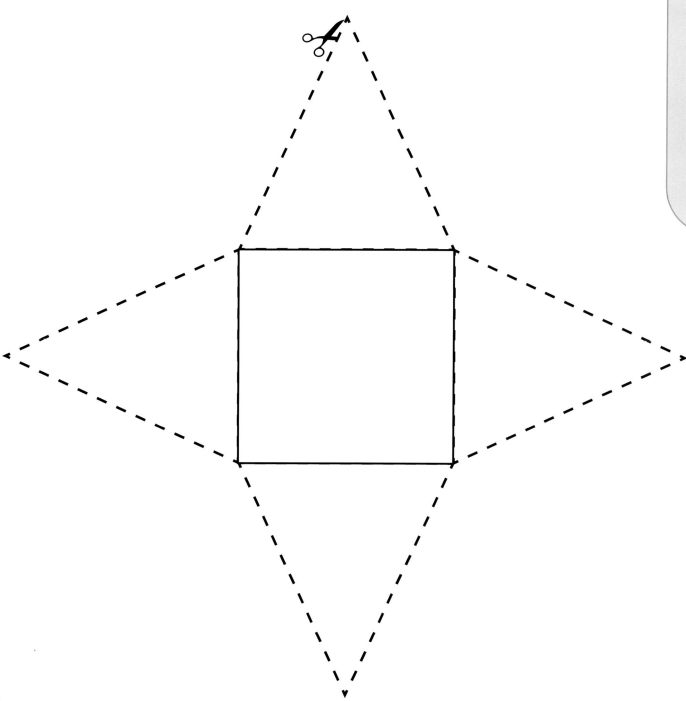

BLM 16   Net Pattern

BLM 17

4-Column Recording Chart

Name _____

Name ───────────────────────────────────

6-Column Recording Chart

Name _____

Directions: Measure the lengths of the lines – first using color tiles, then using centimeter cubes. Write your measurements to the nearest unit.

_____

Length in color tiles: _____ tiles

Length in centimeter cubes: _____ cubes

_____

Length in color tiles: _____ tiles

Length in centimeter cubes: _____ cubes

_____

Length in color tiles: _____ tiles

Length in centimeter cubes: _____ cubes

BLM 19   Line Measurement Worksheet

Glossary of Manipulatives

	AngLegs™ AngLegs enable students to study polygons, perimeter, area, angle measurement, side lengths, and more. The set includes 72 snap-together AngLegs pieces (12 each of six different lengths) and two snap-on View Thru™ protractors.
	Base Ten Blocks Base ten blocks include unit blocks (1 cm on a side), as well as rods that represent 10 units, flats that represent 100 units, and cubes that represent 1,000 units. They can be used to teach number and place-value concepts, such as the use of regrouping in addition and subtraction. They can also be used to teach measurement concepts, such as area and volume.
	Centimeter Cubes These plastic cubes are 1 cm on a side and come in 10 colors. They can be used to teach counting, patterning, and spatial reasoning. They are suitable for measuring area and volume and may also be used to generate data for the study of probability.
	Color Tiles These 1" square plastic tiles come in four different colors: red, blue, yellow, and green. They can be used to explore many mathematical concepts, including those associated with geometry, patterns, and number sense.
	Cuisenaire® Rods Cuisenaire Rods include Rods of 10 different colors, each corresponding to a specific length. White Rods, the shortest, are 1 cm long. Orange Rods, the longest, are 10 cm long. Rods allow students to explore all fundamental math concepts, including addition and patterning, multiplication, division, fractions and decimals, and data analysis.
	Deluxe Rainbow Fraction® Circles The set consists of nine color-coded, 4" plastic circles representing a whole, halves, thirds, fourths, fifths, sixths, eighths, tenths, and twelfths. The circles enable students to explore fractions, fractional equivalences, the fractional components of circle graphs, and more.
	Deluxe Rainbow Fraction® Squares The set consists of nine color-coded, 10 cm plastic squares representing a whole, halves, thirds, fourths, fifths, sixths, eighths, tenths, and twelfths. The squares enable students to explore fractions, fractional equivalences, and more.

	Fraction Tower® Equivalency Cubes Faction Tower Equivalency Cubes snap together to demonstrate fractions, decimals, and percentages. Each tower is divided into stacking cubes that represent a whole, halves, thirds, fourths, fifths, sixths, eighths, tenths, or twelfths. Each cube is labeled with the part of a whole that it represents. One side shows the fraction, another shows the decimal, and a third shows the percentage. The fourth side is blank. Students can turn the cubes or towers to see each of the representations of the same value. Towers, or portions of towers, can be compared with each other.
	Geoboard The double-sided geoboard is 7.5" square and made of plastic. One side has a 5 x 5 peg grid. The other has a circle with a 12-peg radius. Students stretch rubber bands from peg to peg to form geometric shapes. The geoboard can be used to study symmetry, congruency, area, and perimeter.
	GeoReflector™ Mirror This mirror is made of colored, transparent plastic so that the mirror image of an object placed in front of the mirror appears superimposed on the background behind the mirror. The mirror can be used to help students understand transformations, symmetry, and congruence.
	Pattern Blocks Pattern blocks come in six different color–shape varieties: yellow hexagons, red trapezoids, orange squares, green triangles, blue parallelograms, and tan rhombuses. They can be used to teach concepts from all strands of mathematics; for example, algebraic concepts such as patterning and sorting, as well as geometry and measurement concepts such as transformations, symmetry, and area. The blocks can also be used to study number and fraction relationships.
	Polyhedral Dice These dice come in 4-, 6-, 8-, 10-, 12-, and 20-sided varieties and are most typically used for probability activities. They may be used to generate data for number and operations activities and for data analysis.
	Rainbow Fraction® Rings These five plastic rings are used with the Deluxe Rainbow Fraction Circles to make measurements related to circles and fractions of circles. The set consists of a Degree Measurement Ring, a Fraction Measurement Ring, a Decimal Measurement Ring, a Percent Measurement Ring, and a Time Measurement Ring.

	Relational GeoSolids® Relational GeoSolids are 14 three-dimensional shapes that can be used to teach about prisms, pyramids, spheres, cylinders, cones, and hemispheres. GeoSolids facilitate classroom demonstrations and experimentation. The shapes can be filled with water, sand, rice or other materials to give students a concrete framework for the study of volume.
	Snap Cubes® These cubes come in ten colors and can be snapped together to build rectangular prisms and other three-dimensional shapes. They can be used to help students develop spatial sense and learn about surface area and volume.
	Spinners Spinners enable students to study probability and to generate numbers and data lists for number operations and data analysis.
	Two-Color Counters These versatile counters are thicker than most other counters and easy for students to manipulate. They can be used to teach number and operations concepts such as patterning, addition and subtraction, and multiplication and division. Counters can also be used to introduce students to basic ideas of probability.

Index

Boldface page numbers indicate when a manipulative is used in the Try It! activity.

Addition
   Associative Property of, 94
   Commutative Property of, 94, 116
   of decimals, 44
   with Distributive Property, 98
   of fractions with unlike denominators, 40
   Identity Property of, 94
   of integers, 114
   as inverse of subtraction, 106
   of mixed numbers, 34
   in perimeter calculations, 130
Algebra, 92-125
AngLegs™
   Geometry, 60, **61, 63,** 64, **65,** 66, **67,** 68, 70, 72, **73,** 76, **79,** 82, **83,** 84, **85,** 86, **87**
   Measurement, 132, 134
Angles
   classifying, 60
   acute, 60
   obtuse, 60, 62
   right, 60, 62
   straight, 60
   congruent, 84
   corresponding, 84, 86
   measuring, 60, 154
   of a polygon, 66
Area
   of a base, 138, 140
   of a circle, 144
   of a parallelogram, 132, 144
   of a rectangle, 130, 132
   surface, 136
   of a triangle, 134
Associative Property of Addition, 94
Associative Property of Multiplication, 96
Average, 121, 148
Axis
   origin, 123
   x-axis, 70, 74, 122, 123
   y-axis, 70, 74, 122, 123
Base
   of a parallelogram, 132
   of a prism or pyramid, 140
   of a polyhedron, 90, 138
   of a triangle, 134
Base Ten Blocks
   Algebra, 98, 99
   Number and Operations, 26, **27,** 30, 32, 34, **45,** 48, 52, **53**
Centimeter Cubes

Algebra, 94, 104, 110, 114, 116, 118, 120, 122, 123, 124, **125**
   Data Analysis and Probability, **149,** 150, **151,** 152, **153,** 154, 156, **157**
   Geometry, 74
   Measurement, 128, **129,** 130, 132, **133,** 136, 140, 144
   Number and Operations, 22, **23,** 36, 38, **49,** 56
Circle graphs, 154
Circles, 142, 144
   area of, 144
   circumference, 142
Circumference, 142
Classifying
   angles, 60
   quadrilaterals, 64
   triangles, 62
Color Tiles
   Algebra, 96, **101, 103,** 108, 112, **113,** 114, 116
   Data Analysis and Probability, 148, 154, 156, 158, 160, **161**
   Measurement, 128, 130, **131**
   Number and Operations, **37, 39,** 48
Commutative Property of Addition, 94, 116
Commutative Property of Multiplication, 96
Comparing
   decimals, 28, 31
   decimals and fractions, 30
   fractions, 20
   integers, 115
   percents, fractions, and decimals, 32, 34, 154
   percents greater than 100, 34-35,
   predictions and outcomes, 150, 156, 158, 162
   sets of data, 150, 152
Complementary events, 160
Composite numbers, 36
Congruence, 66, 68
Congruent figures
   angles, 84
   corresponding parts of, 84
   figures, 82, 84
   polygons, 84
   transformations, 82
   triangles, 62, 84
Conjectures
   from a data set, 150
Coordinate plane, 122

   origin, 123
   plotting points, 72, 74, 122, 125, 152
   shapes in a, 72, 74
   x-coordinate, 122, 124
   y-coordinate, 122, 124
Corresponding angles, 84, 86
Corresponding sides, 84, 86
Counting Principle
   compound events, 156
   number of outcomes, 156
Cuisenaire® Rods
   Algebra, **95,** 96, **97, 105, 107, 109**
   Number and Operations, 20, 38, 54, **55, 57**
Customary units of measurement, 128
   converting units, 128
   units of area, 130, 132, 134, 136, 144
   units of length, 130, 132, 134, 136, 138, 140
   units of volume, 138, 140, 144
Data
   analyzing, 148
   collecting, 150
   comparing sets of, 148
   graphing, 150, 152
   relationships, 150
   representing
      on circle graphs, 154
      on line graphs, 152
      on scatterplots, 150
Data Analysis and Probability, 146-163
Decimals
   adding and subtracting, 44
   comparing and ordering, 30
   multiplying and dividing, 52
   place value, 26
   relationship to fractions, 28, 32
   relationship to percents, 32, 34
Denominator, 40, 42
Diagonals, 72
Diameter, 142, 144
Dividend, 50
Division
   of decimals, 52
   of fractions, 50
   of integers, 120
   as inverse of multiplication, 48, 108
   meaning, 48
   relationship to ratios, 54

as repeated subtraction, 48
as sharing or equal groups, 48, 52
Divisor, 50
Edges of a polyhedron, 88, 90
Equations, 104
   linear, 122, 124
   nonlinear, 122
   solving addition and subtraction, 106
   solving multiplication and division, 108
Equivalent forms
   of decimals, 28, 34
   of fractions, 24, 28, 34, 40, 42, 50
   of percents, 32, 34
   of proportions, 57
   of ratios, 54, 56
Evaluate expressions, 102
Events
   combined, 162
   complementary, 160
   mutually exclusive, 160
   probability of, 158
   simple compound events, 156, 162
Experimental probability, 158, 162
Expressions, 102
Faces
   of prisms and pyramids, 140
   of a polyhedron, 88, 90
   of a rectangular solid, 137
Factors, 36
   composite numbers, 36
   prime, 36
Flips, 74, 78
Fraction Circles
   Data Analysis and Probability, 154, **155**
   Measurement, **145**
   Number and Operations, **21,** 24, **29,** 30, **33,** 40, **41,** 42, 44, 50
Fraction Circle Rings
   Data Analysis and Probability, 154, **155**
   Measurement, 142
   Number and Operations, 30, **33,** 40, 44
Fraction Squares
   Number and Operations, 20, 24, **25,** 28, 32, 34, **35, 43,** 46, **51**
Fractions
   adding fractions with unlike denominators, 24, 40
   in circle graphs, 154
   comparing and ordering, 20, 30
   equivalent, 24, 28
   dividing, 24, 50
   identifying and writing, 20, 22
   in mixed numbers, 34
   multiplying, 24, 46
   as parts of sets, 22

as parts of a whole, 20
   in probability, 158, 160, 162
   as a quotient, 28
   reasoning with, 20
   relationship to decimals, 28, 30, 32, 34
   relationship to percents, 32, 34, 154
   simplifying, 24
   subtracting fractions with unlike denominators, 24, 42
Fraction Tower® Equivalency Cubes
   Number and Operations, 28, **29,** 30, **31,** 32, 34, 40, 42, 44, 46, **47,** 50
Function
   rule, 110
   tables, 110, 111, 124
   x-values, 110, 122
   y-values, 110, 122
Geoboards
   Geometry, **61,** 62, 64, 70, **71,** 72, 74, 84
   Measurement, 130, **133,** 134, **135**
   Number and Operations, 38
Geometry, 58-91
GeoReflector Mirror™
   Geometry, 68, **69,** 74, **75,** 78, **79**
Graphs
   circle, 154
   four-quadrant, 74, 122,
   line, 152
   linear, 124
   plotting points, 73, 74, 122, 124
Height
   as data, 152
   of a parallelogram, 132
   of a prism or a pyramid, 142
   of a rectangular solid, 136, 138, 140
   of a triangle, 134
Integers
   adding, 114
   comparing and ordering, 113
   dividing, 120
   multiplying, 118
   subtracting, 116
   understanding, 112
Inverse operations, 48, 106, 108, 116, 120
Length, 130, 138
Line graphs, 152
Line segment, 71
Lines
   intersecting, 70
   parallel, 70
   perpendicular, 70
Line symmetry, 68
Line of symmetry, 68
Mean of a data set, 148
Measurement, 126-145
   standard units and precision, 128

Measures of central tendency, 148
Median of a data set, 148
Mental computation
   to solve equations, 104
   using properties, 94, 96, 98
Metric units of measurement, 128
   units of area, 144
   units of length, 138, 140, 144
   units of volume, 138, 140
Mixed numbers, 34
Mode of a data set, 148
Multiplication
   with arrays, 69, 98, 101, 109
   Associative Property of, 96
   Commutative Property of, 96
   of decimals, 52
   with Distributive Property, 98
   factors, 36
   of fractions, 46
   Identity Property of, 96
   of integers, 118
   as inverse of division, 108
   as repeated addition, 46, 52
   in square numbers, 38
Mutually exclusive events, 160
Negative numbers, 112, 114, 116, 118, 120, 122, 124
Nets, 88, 140
Number cubes, 156
Number line, 30, 112, 114, 118
Number and Operations, 18-57
Numbers
   decimals, 26-35, 44, 52
   fractions, 20-25, 28-35, 40-53
   mixed numbers, 34-35
   percents, 32-35
   more than 100, 34-35
Numerator, 40, 42
Operations
   with integers, 114, 116, 118, 120
   relationships between, 46, 48, 52, 106, 108
Ordered pairs, finding on a grid, 72, 74, 122, 124, 152, 153
Order of operations, 100
Orientation, 74, 78, 82
Origin, 123
Outcomes, 156, 161, 162
Parallel lines, 70
Parallelogram,
   area of, 132, 134
   related to the area of a circle, 144
Pattern Blocks
   Algebra, 110, **111**
   Geometry, 64, 66, 68, 69, 76, 77, 78, **81,** 80, 82
   Measurement, 134
Patterns, 110
Percents
   greater than 100, 34

187

    in a circle graphs, 154
    relationship to decimals, 32, 34
    relationship to fractions, 32, 34
Perimeter
    of a rectangle, 130
    of a square, 130
Perpendicular lines, 70
Pi (π), 142, 144
Points
locating on a plane, 72, 74, 122, 124
Polygons
    hexagon, 66, 68, 69, 77, 81
    parallelogram, 64, 132, 144
    pentagon, 66
    octagon, 66
    rectangle, 64, 72
    regular, 66
    rhombus, 64, 72
    square, 66, 72
    trapezoid, 64
    triangle, 66, 76
Polyhedrons, 88
Polyhedra Dice
    Algebra, 102
    Data Analysis and Probability, 158, 160, **163**
Positive numbers, 112, 114, 116, 118, 120, 122, 124
Predictions,
    making, 66, 110, 150, 152, 158, 162
Prime factor, 36
Prime factorization, 36
Prime numbers, 36, 158, 160
Prism
    cube, 90
    rectangular, 88, 90, 138
    triangular, 140
    volume of, 140
Probability, 156
    complementary events, 160
    counting principle, 156
    of an event, 158
    experimental, 158, 162
    of mutually exclusive events, 160
    of simple compound events, 162
    theoretical, 158, 162
Products, 46
Proportions, 56
Protractor, 60, **61, 63,** 66, **81**
Pyramid
    square, 140
    triangular, 90, 140
    volume of, 140
Quadrant, 122
Quadrilateral, 64
    classifying, 64
    parallelogram, 64, 132, 144
    rectangle, 64, 132
    rhombus, 64, 66
    square, 64, 66

    trapezoid, 64, 66
Quotient, 28
Radius, 144
Range of a data set, 148
Ratio(s)
    of diameter to circumference, 142
    finding equivalent, 54, 56
    meaning of, 54
    relationship to proportions, 56
Reciprocal, 50
Rectangular solids
    surface area of, 136
    volume of, 138
Reflections (flips), 68, 74, 78, 82
Regular polygons, 66
Relational GeoSolids®
    Geometry, 90, **91,** 122, **123,**
    Measurement, 138, 140, **141,** 142, **143,** 144
Rotational symmetry, 76, 80
Rotations (turns), 68, 76, 78, 82
Sampling
    to make conjectures, 150
Scatterplots, 150
Shapes
    changing orientation of, 74, 78
    in a coordinate plane, 72, 74
    relating two- and three-dimensional, 88
    three-dimensional, 88, 90
    two-dimensional, 88, 90
Sides
    of an angle, 60
    congruent, 84
    corresponding, 84, 86
    of a polygon, 66
    of a triangle, 84
Sieve of Erathosthenes, 36
Similar
    figures, 84
    triangles, 62, 86
Similarity, 66, 68, 86
Simple compound events, 162
Slides, 74, 78
Snap Cubes®
    Algebra, , 94, 96, 100, 106
    Measurement, 136, **137,** 138, **139**
    Number and Operations, 36, 44
Spatial relationships, 70
Spatial visualization, 88
Spinners
    Data Analysis and Probability, 156, 158, **159,** 160, 162
    Measurement, 142
Square, 66, 72
Square numbers, 38
Square roots, 38
Standard units and precision, 128
Subtraction
    of decimals, 44

    of fractions with unlike denominators, 42
    of integers, 116
    as inverse of addition, 108
Surface area
    of a rectangular solid, 136
Symmetry, 68, 76, 80
Tangrams, 132
Tessellations, 80
Theoretical probability, 158, 162
Tiling patterns, 80
Transformations, 68, 74, 76, 78, 82
    applying and describing, 68, 76, 78, 80
    and congruent figures, 82
    multiple, 78, 82
    reflections (flips), 68, 74, 78, 82
    rotations (turns), 68, 76, 82
    translations (slides), 68, 74, 78, 82
Translations (slides), 68, 74, 78
Triangles
    area of, 134
    classifying, 62,
        acute, 62, 86
        equilateral, 62, 66, 86
        isosceles, 62, 84, 86
        obtuse, 62
        right, 63
        scalene, 62, 86
        congruent, 62
        number in a polygon, 66
        similar, 62
Turns, 76, 78
Three-dimensional shapes, 88, 90
Two-Color Counters
    Algebra, 98, 100, 102, 106, 104, 108, 112, 114, **115,** 116, **117,** 118, **119,** 120, **121**
    Data Analysis and Probability, 148, 152, 156, **157, 163**
    Measurement, 142
    Number and Operations, 22, 54, 56
Two-dimensional shapes, 88, 90
Variable, 102
Variable
    expressions, 102
    equations with, 104
Vertex
    of an angle, 60, 80
Vertices
    of a polygon, 66, 74
    of a polyhedron, 88, 90
Volume
    of prisms, 140
    of pyramids, 140
    of a rectangular solid, 138